JANUA LINGUARUM

STUDIA MEMORIAE
NICOLAI VAN WIJK DEDICATA

edenda curat

C. H. VAN SCHOONEVELD

Indiana University

Series Minor, 133

COLLOQUIUM PAEDOLINGUISTICUM

Proceedings of the First International Symposium of Paedolinguistics

Held at Brno, 14-16 October 1970

edited by

KAREL OHNESORG

1972

MOUTON

THE HAGUE · PARIS

LIBRARY OF CONGRESS CATALOG CARD NUMBER: 72-86849

Printed in Belgium by NICI, Ghent

PRÉFACE

Nous présentons les rapports sur le premier colloque international de pédolinguistique, organisé à Brno (Tchécoslovaquie) du 14 au 16 octobre 1970.

Le colloque fut organisé par l'Association linguistique de Brno, affiliée à l'Académie des Sciences, et par la Faculté des Lettres de l'Université de Brno, sous l'égide de l'International Society of Phonetic Sciences.

Le Comité d'organisation adressa en mars aux chercheurs qui pourraient s'intéresser au sujet proposé des invitations avec la prière de bien vouloir indiquer les adresses d'autres chercheurs dans leurs pays qui voudraient participer au colloque. Le symposium fut aussi annoncé, au mois de mai, dans le circulaire N° 10 de l'ISPhS. Jusqu'au premier octobre 60 participants de 14 pays s'inscrivirent. Participèrent ensuite 48 chercheurs qui firent 28 communications suivies de 62 discussions enregistrées.

Le colloque fut inauguré dans l'aula de la Faculté des Lettres. Le symposium ayant été ouvert par son président le Prof. Karel Ohnesorg, le Prof. Martin Kloster-Jensen fit une allocution en tant que Secrétaire général de l'Association internationale des Sciences phonétiques. Souhaitèrent ensuite pleins succès au congrès le Prof. Roman Mrázek (désigné par le doyen de la faculté), le Dr Helmut Stelzig (représentant de l'Université de Greifswald) et le Dr Antonín Vašek (secrétaire de l'Association linguistique de Brno).

Le secrétaire du colloque, le Dr Lubomír Bartoš, lut ensuite le texte des télégrammes adressés aux pionniers de la pédolinguistique, au Prof. Marcel Cohen (Paris) et au Prof. M.E. Chvatcev (Léningrad). Le troisième de ces pionniers, le Prof. Milivoie Pavlović étant présent, le président lui rendit hommage personnellement.

Suivirent les séances plénières (Mme T. Slama-Cazacu, Mme Suzanne Borel-Maisonny, M. Milivoie Pavlović). Les autres communications et discussions eurent lieu dans deux sections, présidées alternativement par S. Borel-Maisonny, L. Kaczmarek, M. Pavlović, V. Rūke-Draviņa, T. Slama-Cazacu et H. Stelzig.

La clôture du colloque eut lieu en séance plénière où sur proposition du Prof. M. Pavlović, il fut décidé de charger l'Institut de Phonétique de Brno de l'organisation des études pédolinguistiques dans le cadre des activités internationales. M. Kloster-Jensen ayant souhaité, dans son allocution, aux recherches pédolinguistiques un essor favorable, le président du colloque remercia tous les participants du symposium pour leur assistance et clôtura le colloque par un épilogue latin. — Le 19 octobre, M. Kloster-Jensen rendit compte du colloque à l'UNESCO, à Paris.

Nous publions les textes des communications de notre colloque. Comme quelques auteurs, pendant la rédaction définitive du texte, ont dépassé l'étendue limitée, en ne voulant point restreindre leurs manuscrits, nous renonçons à la publication, dans le présent volume, des résumés rédigés dans une autre langue que celle de la communication. — Cependant, ces résumés ont paru, en 1971, dans la revue *Sborník prací filosofické fakulty brněnské university — Studia minora facultatis philosophicae Universitatis Brunensis* (*SPFFBU* A 19 [1971] 215-227).

Brno, décembre 1970

KAREL OHNESORG
le Président du colloque

LUBOMÍR BARTOŠ
le Secrétaire du colloque

CONTENTS

(This volume contains the texts which had been received by the editor by December 31 st, 1970)

Préface 5
Prologus Colloquii Praesidis 11
Inaugural address by Prof. Dr. M. Kloster-Jensen 13

PAPERS OF PLENARY SESSIONS

T. Slama-Cazacu.
Fifty Years of European Child Language Studies, and Perspective in this Field 17
S. Borel-Maisonny.
Tomoanalyse et oscillographie 37
M. Pavlović.
Trois principes concernant le phonème 59

PAPERS OF SECTION MEETINGS

L. Bartoš.
La réalisation des groupes consonantiques chez un enfant tchèque 71
J. Dvončová.
Das Problem der Beziehung zwischen der Artikulationsbasis und dem phonologischen System in der Erforschung der Kindersprache 75

M. Chmura-Klekotowa.
Словообразовательные неологизмы в детской речи на материалах полского и русского языка 83
P. Janota.
Development of Children's Vocabulary 101
N. Janotová.
Gebrauch der Eigenschaftswörter bei hörgeschädigten Kindern 110
L. Kaczmarek.
Sprachanbildung bei der gehörlosen blinden Helene Keller 115
J. Kania.
Phonetik der Kinder mit verzögerter Sprachentwicklung 124
W. Kaper.
Normalsprachige Transformationsgeschichten und kindliche Satzbildung 129
M. Kloster-Jensen.
La constatation d'oppositions phonémiques dans le système linguistique de l'enfant 139
Vl. Lejska.
Einfluss einiger wichtigen Gesellschaftsfaktoren auf die Entwicklung der richtigen Artikulation der Kinder . . 144
J. Liška.
Articulatory Disorders and their Effect in Community . 152
Klára S. Meggyes.
The Synchrony of Different Developmental Stages in the Language of a Two-Year-Old Child 159
M. Mikeš, L. Dezsö, P. Vlahović.
Sentence-Programming Span in Child Language . . . 166
N. Milan.
A Study of Children's Speech in a Spontaneous and Non-Spontaneous Situation 179
K. Ohnesorg.
Une contribution à la pédophonétique comparée . . . 187
E. Oksaar.
Zum Erwerb der estnischen Quantitätsregeln im zweisprachigen Milieu 193

J. Pačesová.
The Growth of Phonemic Repertory in Czech-Speaking Children 199
J. Průcha.
On the Psycholinguistic Study of Language Acquisition 208
V. Rūķe-Draviņa.
The Emergence of Affirmation and Negation in Child Language 221
G.W. Shugar.
A Method of Descriptive Analysis of Child Activity in the Period of Language Acquisition 242
M. Sovák.
On Linguistic Sense from the Point of View of the Reflex Theory 254
M. Vajshajtlová.
Das Gefühl und die Sprache 256
P. Vlahović, M. Mikeš, L. Dezsö.
Développement des constructions de complément d'objet et de lieu dans le langage des enfants serbocroates et hongrois. 260
M. Zarębina.
Disorders of Speech in Child Aphasia 270

Closing Session

M. Kloster-Jensen 277
K. Ohnesorg
Epilogus 278

Colloquii participes 280

Colloquii inauguratio

PROLOGUS

COLLOQUII PRAESIDIS

Feminae doctissimae, viri doctissimi, cari sodales,

Permittere velitis quaeso, ut vos salutem in urbe nostra, ubi convenimus, ut studio linguae infantium operam daremus. Quae studia hoc saeculo cum tanto successu excoluntur, ut origines disciplinae novae, paedolinguisticae, immo vero paedophoneticae quoque, exoriri videamus. Neque mirum est. Nam indagatio linguae infantium, ad praecepta linguisticae diligentissime redacta, multos aspectus offert, quibus nonnullae considerationes de linguarum ratione ac evolutione illustrantur, complentur, confirmantur. Magnopere autem gaudeo, quod huius generis symposium primum in patria nostra instituitur, in regione, ubi iam tribus saeculis ante Comenius ille rationem studii huius quaestionis habuit, praecipue in libro, cui Informatorium scholae materni gremii inscribitur. Itaque iustum et aequum habemus nomen auctoris huius, trecentos annos post eius obitum, inducere ac in hoc consessu grata memoria prosequi.

Salvere vos iubeo omnes, qui e quattuordecim regionibus, complures per multas gentes ac immo vero per aequora vecti, ad symposium nostrum convenistis, et plenum successum laboribus, disputationibus colloquiisque vestris ex corde et ex animo exopto.

Quod felix, faustum fortunatumque eveniat.

INAUGURAL ADDRESS BY THE SECRETARY GENERAL OF THE INTERNATIONAL SOCIETY OF PHONETIC SCIENCES, PROF. DR. MARTIN KLOSTER-JENSEN

Mr. President, Dear Colleagues,

It is with great pleasure and gratitude that I have accepted the honour to salute this symposium on behalf of the International Society of Phonetic Sciences. It is also my privilege to thank the President, Professor Karel Ohnesorg, for the strenuous work which he is ending successfully today by starting the series of papers to be read, with discussions to follow, in a field which is of prime importance to the advancement of linguistics and therefore to several branches of humaniora, namely that of child language. This part of linguistics does not seem yet to have received all the attention it deserves. It is on this basis that our President has taken the welcome step to create, for three days to begin with, an atmosphere favourable to the solution of problems in the exclusive domain of child language.

Mr. President, you have chosen the moment extremely well. The whole world is celebrating the memory of your famous educator Comenius, and this year is also devoted by UNESCO to the extension of the foundations of teaching and creative ideas on education, with the symbol IEY: International Education Year. Teachers and researchers in all countries have been asked with all possible engagement by the Director-General of the United Nations Educational, Scientific and Cultural Organization to give a new impetus to the advancement of learning, training, and education. The International Society of Phonetic Sciences is happy to support every such arrangement, and is indeed proud to be able to patronize this symposium.

May I also express the hope that our work in Brno will prove useful not only for our research projects, but that it will also add important arguments to a plea for the great idea conceived by Professor Ohnesorg: to establish in different universities chairs for the study, research and teaching in the field of child language.

PAPERS OF PLENARY SESSIONS

FIFTY YEARS OF EUROPEAN CHILD LANGUAGE STUDIES, AND PERSPECTIVE IN THIS FIELD

TATIANA SLAMA-CAZACU

(Bucureşti)

1. A retrospective view on European child language studies during the last fifty years is gladdening and encouraging. Numerous studies — some of them very ample or written by well-known specialists in their area (or who arrived to notoriety through these works) — have demonstrated the importance of child language research, not only for psychology or pedagogy, but also for other fields, like linguistics, phonetics, neurology, defectology, or even cybernetics.

1.1. Actually it is hard to say if child language studies 'pertain to' or are 'of interest for' only psychology or only linguistics (speaking now about the fundamental research and not about the applied fields); perhaps, in reality, child language represents a common topic for both sciences and in fact it should be approached by an interdisciplinary methodology or, at any rate, taking into account the perspective of both linguistics and psychology. Anyhow, this study has been for many years especially performed by psychologists, who, even if they sometimes tried to collaborate with linguists, lacked however a special training in order to observe and record facts in a satisfactory manner from the linguistic point of view. However, it is important to mention here not only the interest of well-known psychologists of the twenties or early thirties for the language of children, but also their effort to attain to a linguistic understanding of facts, at least studying the works of linguists, collaborating with them in collective volumes, or participating at meetings where both linguists and psychologists were represented, and even to linguistic societies (as, for instance, the meetings of the Société linguistique de Paris, where we sometimes find together the names of to A. Meillet, M. Cohen and H. Delacroix, H. Piéron, H. Wallon, O. Bloch). Psychologists have been for longer time fascinated by the study of child language and this science, nowadays, still is illustrated by a great number of passionate investigators of the field. Let me mention here the typical attitude of one of them (who is revendicated by pedagogy but also by psychology), Ovide Decroly [1933]: his wife wrote on the copy of the book *Comment l'enfant arrive à parler* (posthumously published in 1936 and which she offered to H. Piéron) that it was a work "auquel mon mari a consacré encore la dernière soirée de sa vie, et cela avec tant de joie et d'animation" (Library of the "Institut de psychologie", Paris).

We may say that especially around the twenties outstanding linguists not only became more aware of the interest of this topic, but dedicated even ample studies — sometimes volumes — to it. Some of them underlined the importance of this field for linguistics, as O. Jespersen, who dedicated to it one chapter of his book on *Language* (1922), writing that "this study seems to me very fascinatory indeed, and a linguist is sure to notice many things that would be passed as uninteresting even by the closest observer among psychologists, but which may have some bearing on the life and development of language". An outstanding linguist who is well-known for his pioneering book on early bilingualism, M. Pavlovitch (1920), also insisted on this idea, writing, in 1920, that this is "a study indispensable for linguistics". Linguistic studies were however not very numerous, and this explains perhaps the insistence of some great linguists to advocate for them. Somewhat later, R. Jakobson ([1939], [1941], etc.) would demonstrate this fact through his works and would even apply and verify the new principles of phonology in investigating child language; "the conclusion was that child language — much the same as aphasia — might and ought to be incorporated into general linguistics" (Grégoire, 1949-1950: 356). It is impressing to know that the first attempt to apply the new structural linguistic methodology was made on child language and that also in this early work of Roman Jakobson one can see the part played by *Gestalt*-psychologists in the appearance of linguistic structuralism. J. Vendryes (1953: 27), on reviewing one of M. Cohen's (1952) articles, concluded that "L'étude du langage enfantin est pour le linguiste d'une importance capitale", because the path followed by mankind may be observed "en raccourci" on the child. Cohen himself has maintained the usefulness of these studies for general linguistics, both by the methodology of some of his works — in which facts from child language are confronted with linguistic phenomena noticed in various languages — and by repeated assertions (requesting, among other things, even the setting up of chairs for the study of child language — Cohen, 1951: 65). Concerning the opinion expressed by Cohen (1952), a reviewer of his article wrote that: on the basis of an "enormous amount of facts, scientifically established and verified, one could obtain, from child language, precious knowledge for general linguistics" (Wiblé, 1954: 75). One of A. Meillet's disciples, A. Grégoire (1949-1950), author of one of the best books on child language, using a rigorous linguist's method, asserted (taking up one of A. Busemann's formulae applied to Grégoire in 1937) that "the scientific revival of 'child-linguistics' " ("linguistique enfantine") had begun. This idea of a 'revival' was later on taken up by K. Ohnesorg too (1955b: 95; 1962), who even proposed the term 'pedophonetics' or 'pedolinguistics' (Ohnesorg, 1955b: 95), concluding that "pedolinguistics may substantially contribute to the study in general linguistics" (Ohnesorg, 1955b: 101; 1959: 149; 1964a; 1964b).

These ideas have become current nowadays; however, truth found but gradually its way into the mind of a great number of scientists, until it became characteristic of our epoch. The utility of child language study for general linguistics has been also officially recognized, by creating a section devoted to this field in the frame of the Xth International Congress of Linguists (Bucharest, 1967). The impetus has got hold also of the researchers across the Ocean, where child language investigation — at present mostly through the generative grammars methodology — has become a vogue, that — we hope — will probably also lead to a real and thorough understanding of child language features.

Yet, to reach this point, much time had to elapse. It is fascinating to analyse this path — covered by attempts and enthusiasm for investigation, by success and jeopardized hypotheses, by the tendency to accumulate facts or perhaps, on the contrary, towards premature generalizations. It is a path characteristic to European scientists, as American researchers — particularly the linguists (the bulk of investigators as well as their outstanding exponents) — have been won over to this study somewhat later.

2. The fact that child language is generally worth studying became evident (and this evidence got materialized in the very first attempts of investigation) particularly towards the end of the European XIXth century.

2.1. As it always happens, in the history of sciences, some sporadical preoccupations existed also before, and the first name that can be quoted is (as Leopold, 1959: 191 affirms) that of D. Tiedemann, in 1787. Let us also mention, among the pioneers of this field, also Baudouin de Courtenay, with a study of 1870 (cf. Leopold, 1959: 209).

Somewhat later, importance begun to be attached to the study of child language also with respect to what it might yield for a deeper understanding of the general mechanisms of language, or for the general characteristics of the 'language' (Fr. *langue*) phenomenon itself. The observation of one's own child was the most frequent device: certain researchers were attracted by bilingualism (such as J. Ronjat 1913) — a promotor in this field, without whom A. Grégoire would be more difficult to be understood — or by language facts from a less known linguistic area (I. Georgov, 1905; 1908; 1912; cf. Leopold, 1952: 3, for the Bulgarian, for instance; or E. Tappolet, 1907, *ibid.*: 94-95, for the Basel dialect, etc.). A number of studies have been the output of the practical interest in speech troubles, in the re-education of deaf-and-dumb children, in the severe pathology of the aphasics (H. Gutzmann, [1931, 2nd ed.; 1894], etc.).

2.2. In the twenties, shortly after the first World War, important works came into circulation — which were however either initiated during the pre-war period or bearing the mark of previous formation —, or reprints of some older books (as for instance those of Van Ginneken, *The Novel of a Child* [1917] and *The Discoveries of a Child* [1924, in Dutch], or O. Jespersen, *Child Language* [1923, in Danish], or C. and W. Stern, *Kindersprache* [1928, new edition], etc.). An increased interest is taken in cases of bilingualism (first M. Pavlovitch, [1920]; then others, such as A. Grégoire, [1933] — whose works would appear also later, in 1937; 1947), making use of a linguistic method, sometimes of phonetic transcriptions (as in A. Grégoire's excellent monograph).

In general, the same procedure of observing certain 'cases' is used (and these are mostly provided by the linguist's own children). But studies of synthesis are not missing either, which are discussing certain principles or some general tendencies of the evolution of child language, and which are worked out on the basis of biblio-graphical data or on the author's own observations: from O. Jespersen to H. Delacroix (one of his articles, "Au seuil du langage" [1933], and a book, *L'enfant et le langage* [1934]), and from K. Bühler (1926; 1930) or Ch. Bühler (1930), to O. Decroly (about the year 1930).

But, still in the early twenties, certain quite new preoccupations became evident and inspired even some of the works that had been started before. Besides 'inventory' monographs in which the descriptive effort or interpretations belonging to outdated phases in linguistics or psychology prevailed (as in Clara and W. Stern's book), besides works of mere psychological analyses or of discussions on an exclusively speculative psychological level (concerning imitation, for instance), besides works of a psycho-pedagogical interest, we find also works in which the old question of the logico-grammatical parallelism is tackled through new procedures or by more subtle interpretations of the relationship between thought and language (an old *leit-motif* in the works of J. Piaget, in the excellent articles of O. Bloch, [1921; 1923; 1924]; and of P. Guillaume [1927a; 1927b]; in H. Delacroix, [1924b]; in L. Vygotskij, see the [1956] ed., a.s.o. — that has come down to this day without turning obsolete). Elaborated chiefly by psychologists, these works are not always very concrete, and sometimes quite far from being analytical studies of language — with the exception, perhaps, of Bloch's and Guillaume's.

There had been, of course, also certain psychologists who noticed the necessity of approaching the linguistic aspects or of using a method to suit these aspects. But we can easily trace in their attempts the limits of self-formation in this field (as in the case of the Sterns). The necessity was met by the collaboration, on an utterly exterior level — i.e. by means of joint circles where papers were delivered or by publishing articles written by psychologists but also by linguists, neurologists, etc., in collective volumes (such as the special issues of the French *Journal de psychologie*: the issue of 1931 has become a 'classic', under the name of *Psychologie du langage* [1933]). The linguists, on the other hand, cannot either give a sufficiently thorough interpretation of the facts from the point of view of the psychological age-peculiarities of the child, and the exposition of their results often incites questions as to the generalizing value of the data gathered by observing one or several 'cases' (as, for instance, Grégoire's monograph *L'apprentissage du langage* I [1937], remarkable by its method of minute linguistic analysis and for its wealth of material — collected, however, from his own two children only).

2.3. The inceptive innovation in the linguistic analysis of child language (with bearings also upon the physiological and pathological plane) was marked by R. Jakobson's first publications [1939] [1941] in this field (especially his book, of 1941, *Kindersprache, Aphasie und allgemeine Lautgesetze*). It is by him (much more than A. Grégoire thought — see the second volume of his monograph, [1947: 345] — he had done it himself), that the method of phonological analysis has been introduced in the study of child language; in fact, it is the structuralist trend that penetrates thus here too (much more than through *Gestalt*-psychology). But the ideas and analysis suggested at that moment would not yield fruits at once: the more so as the second World War was just breaking out.

2.4. After the second War, the upsurge was manifested chiefly by the linguists, some of them following the path opened by R. Jakobson. Around 1950 and next to the end of the decade, minute studies came out: A. Grégoire (1947 — the final part of his monograph), A. Gvozdev (*Formirovanie u rebenka grammatičeskogo stroja russkogo jazyka* [1949]), L. Kaczmarek (*Kształtowanie się mowy dziecka* [1953]), K. Ohnesorg (*Fonetická studie o dětské řeči* [1948], and many other studies).

Some psychologists, linguists, or pedagogues either keep dwelling on older problems, or approach them from the same old points of view, or evince more marked preoccupations for the aspects concerning practice (i.e. studies on the language of the deaf-mutes: P. Oléron (*Recherches sur le développement mental des sourds-muets* [1957], and other studies), B. Tervoort (*Structural Analysis of Visual Language Use with a Group of Deaf Children* [1953, in Dutch], and other studies). The relation between language and the whole development of the child is part of the same psycho-educational preoccupation (A. Luria, *Rol' slova v formirovanii vremennyx svjazej v normal'nom i anormal'nom razvitii* [1955], and other studies), which is particular especially to the Soviet school (where, for a time, the Pavlovist conception about the nervous high activity and the development of the two signalling systems had also been manifest).

The methodological lines represented by psychology on the one side, and linguistics on the other, keep still apart, with certain exceptions. Some research-workers (let us mention here T. Slama-Cazacu, studies between 1956 and 1968) — aware of the necessity of both approaches, or rather of the necessity of combining them, as well as of replacing the casuistic method by studies carried out on a lot of children, that should allow generalizations — do try to surpass both the operation of mere linguistic inventorying and the exclusively speculative psychological analysis

(*Relations between Thought and Language in Ontogenesis* [1957, in Romanian], and other studies). The above mentioned author published works based on such an approach, in the domain of the relationship between thought and language, or less studied themes are looked for, such as children's dialogue, the style of child speech, etc. The analysis effected on such bases coincided also with the outspreading of the term 'psycholinguistics' in the U.S. (see Osgood and Sebeok, 1954).

Gradually, the new linguistic methodology begins to yield results, the proper linguistic studies contain more accurate notations, make use of a more firm method of analysis or broach new topics (e.g. K. Ohnesorg's second monograph, *Druhá fonetická studie o dětské řeči* [1959]; W. Kaper's studies, *Kindersprachforschung mit Hilfe des Kindes* [1959], and other studies; V. Rūķe-Draviņa, *Zur Sprachentwicklung bei Kleinkindern: Syntax* [1963], *Mehrsprachigkeit im Vorschulalter* [1967], and other studies); structural analyses appear, and the influence of the theory of information or of the generative grammar makes itself felt.

2.5. One may say that in the 60's a smaller number of important studies of detail appeared in Europe than across the Atlantic (yet among these studies there are vaster and more remarkable monographs, as for instance those published — from among the linguists — by V. Rūķe-Draviņa). Although about 1960 the opinion had been expressed (Leopold, 1959: 193) that the moment was not yet ripe for a general view and that it was good to keep on effecting detailed studies and gathering on material (somehow on 'cases'), nevertheless the procedure of collecting data changed somewhat (the more so since the increased use of the tape- recorder and the progress registered by the methodology of applied linguistics made it possible and compelling). Let us mention here also the more recent book — containing especially a bibliographical discussion on some works, in the field — of G. Francescato, *Il linguaggio infantile* [1970], the works of S. Borel-Maisonny (1960, and others), of W. von Raffler-Engel (1964, and others), on bilingualism of A. Tabouret-Keller (1963; and in M. Cohen *et al.*, 1962). To the investigation carried out on a great number of children — upon resorting to the psycholinguistic methodology as well as to the technical devices offered by portable tape-recorders — and to studying new aspects, as for instance children's dialogue (Slama-Cazacu, 1961; 1966a; 1968), endeavours are added to do complex research, in teams, on a great number of subjects. Projects are drawn up for the elaboration of corpora relative to English, French, etc., or for international co-operation (as that one which connected, in 1963, Czech, Italian, Polish, Romanian researchers: G. Francescato, L. Kaczmarek, K. Ohnesorg, T. Slama-Cazacu), in order to study the formation of the child's phonemic system (see Slama-Cazacu, 1966b; 1967), trying to get a thorough knowledge of the child's vocal emissions during the first year of life. Some of these works are conceived along the modern linguistic principles, and sometimes along the principles of the generative grammar (yet, in this last case not always creatively, but rather as a simple application of the American experience in generative grammars or of a certain method connected with the name of a certain author).

In their turn, the psychologists begin initiating themselves into the modern linguistic conceptions and attempt to put into operation the methods of linguistic analysis. However, until now it cannot be asserted that in Europe the psychologists have succeeded in working out original studies on child language by using these linguistic methods. And on the other hand, the proper psycholinguistic integration still is at an incipient stage in which, for many European psychologists, the proper object and the aims of psycholinguistics are not clearly defined.

2.6. The modern concern for an interdisciplinary approach or from multiple angles is reflected in the study of child language too; to these one may add the tradition of co-operation in the field of language or generally on communication, of certain psychologists, linguists, neurologists, sociologists, 'defectologists', technicians, etc. Hence, the publication of collective volumes, which deal with child language from several standpoints (e.g. *Bulletin de psychologie* 19: 8-12, [1966], under the editorship of P. Oléron), or international meetings (as the XVIIIth International Congress of Scientifical Psychology, Moscow, 1966; or the Xth International Congress of Linguists, Bucharest, 1967; the VIth International Congress of Phonetical Sciences, Prague, 1967, etc.), where child language is being discussed upon by specialists in various branches, and now this first European Conference on 'Pedolinguistics' — in fact, on child language studies — owing to the enthusiasm and competence of one of the promotors of child linguistics in Europe, Karel Ohnesorg.

The practical need made research develop in educational psychology (on the teaching in schools of the vernacular or of foreign languages, of reading and writing, on the oral and written language of children, etc), or logopedics and 'defectology' (mention must be made in this respect of the special logopedic and 'defectologic' schools in the Soviet Union, Czechoslovakia, the Netherlands, Poland, etc.).

3. As far as the linguistic geographic map of Europe is concerned — from the point of view of the 'sources' or of inventorying the various forms and peculiarities connected with the respective language —, one may say that it is getting filled in. To previous studies, for English, French and German, and the sporadic ones for Bulgarian (I. Georgov), Danish (O. Jespersen) or Dutch (J. Van Ginneken) for instance, new ones have been added after the first World War, containing material on: Danish (O. Jespersen, 1923), Dutch (J. Van Ginneken, 1924), Hungarian (E. Kenyeres, 1926; 1927; etc.), Polish (S. Brenstiern-Pfanhauser, 1930), Russian (Rybnikov, 1926; 1927; and others), Serbian (M. Pavlovitch, 1920), and others for English (M. Lewis, 1936, etc.), French (O. Bloch, 1921; 1923; 1924; M. Cohen, 1925; 1933; O. Decroly [in the thirties]; A. Grégoire, 1933; 1937, 1947; and others), German (C. and W. Stern 4th ed. 1928, and others). And after the second World War, containing material on: Czech (particularly K. Ohnesorg, 1948; 1955a; 1959), Dutch (W. Kaper, 1959; 1967), Finnish-Swedish (B. Malmberg, 1945), French (M. Cohen *et al.*, 1952; 1962; A. Grégoire, 1937; 1947; 1948; and others), Italian (P. Fronzaroli, 1957; W. von Raffler-Engel, 1964, etc., G. Fran-

cescato, 1970), Lettish (V. Rūķe-Draviņa, 1963; 1967; etc.), Polish (L. Kaczmarek, 1953, and other psychologists and linguists), Romanian (T. Slama-Cazacu, 1957a; 1957b; 1961; 1962a; 1962b; 1968; etc.), Russian (A. Gvozdev, 1949; K. Čukovskij, 1955; and others), Serbo-Croatian (D. Kostić and M. Stosić, 1963; M. Vasić, 1965; etc.; M. Mikeš, 1964; 1967; etc.).

4. My own impression is that we are living now a period of 'expectation', wherefrom important studies will appear in the years to come. The decade of the 70's will probably offer to those who will live throughout it new and original works, resulting very probably from the confluence of psycholinguistics, new structural and generative linguistics, Piagetian psychology, present-day social psychology and the trends of applied linguistics (especially in the fields of language pathology, foreign-language teaching, communication improvement in grown-ups, and the mass-communication media and their part in modern education).

If I have begun my paper by expressing a satisfaction when contemplating the result of these fifty years of human struggle to come to know an important side of what makes man be man — I mean to know how a human being begins to speak —, I have now to lucidly point up the shortcomings of an image which might appear too much as an Arcadian painting made in idyllic colours. One must realize that we still know little about the development of child language, about its stages, its general features, about the part played by native background and by social impact, about the particuliarities this development has in common for all children independently of the language community in which a child is born.

We would desire, first of all, to see that the European map has not any more 'white regions', 'terra incognita', we know still very little about child language in Finnish, Hungarian, Norwegian, Swedish and other languages as well, and almost nothing about Albanian, Estonian, Greek, Portuguese, Spanish. There is the need for more AMPLE RESEARCH, considering not only isolated cases, but on a GREAT NUMBER OF CHILDREN for every language (as also M. Cohen pointed out in 1952: 222: "un grand nombre d'enfants devront être observés, en tous pays, de manière à reconnaître les règles d'acquisition").

There are still lacking studies — in diverse languages and giving the possibility to objective comparative analyses — concerning the development of phonematic system in the early period of life. There are not sufficient studies on the development of GRAMMATICAL STRUCTURES IN DIFFERENT LANGUAGES, based on a reciprocal integration of linguistics and psychology, and which would be not merely inventories of forms, but neither a discussion on pure cognitive aspects. We still need more and more studies on the DIALOGUE among children in different languages, on NON-VERBAL ASPECTS — which play such an important part, intermingled in the verbal ones, generally in communication and especially in that among children —, and also studies based on a methodology taking into account all the situation — the context in broad and more restricted sense — in which a child language form is noted. One needs to find out the objective and refined means for an efficient COMPARISON BETWEEN DIFFERENT LANGUAGES, for phonematical, grammatical, semantical evolution, in order to establish what is UNIVERSAL, at least in these few languages of the world that are spoken in our old Europe. In this concern, one needs to establish what is 'born' (defining however this concept of innateness and delimitating linguistic features from the cognitive aspects or from the biological background that is, of course, born in the sense of a 'species' heredity, namely of 'being born as a human being'). In order to arrive at such an end, one still needs, however, to show the part played by social influence, by the perception of a language spoken by those who surround a child, and also to delineate the role of this mechanism — still unknown even by psychologists — of 'imitation', in the context of a reasonable understanding of maturation and generally of the active part of the child, who — as a selective echo — filters through his capacities whatever society tries to impose on him.

Generally, studies concentrated on details are expected to be more and more replaced by vast syntheses, based on a great amount of facts, collected by the authors themselves. Let me also hope that a CREATIVE elaboration of psycholinguistics (and not merely a replication of theories in fashion), in an original European form, should be taken more and more as a theoretical and methodological background for these studies to come.

I would like to conclude by a challenging question, concerning the attraction child language has exerted on so many scientists and by another one, on the glimpses of the future research, which can already be seen in some cues of present-day studies.

What made Jespersen and Piaget, Vendryes and Luria, Van Ginneken and Jakobson, Grégoire and M. Cohen, and so many other research-workers not less renowned or less important in the history of the respective sciences, what made all these scientists abandon for a while their direct or permanent research object, and bend — in the proper sense of the word — over these unassuming subjects, from whom, nevertheless, one always learns more than out of books? What made other European specialists dedicate almost all their activity to the same study — of cries, of mumbling apparently nonsenseful, of the often cryptic galimatias, and then of the chattering that gradually gives way to the not seldom sophisticated and refined language of the adolescents? Was it the pleasure of getting in touch with these little creatures full of mystery? Was it the charm of discovering less known aspects, that also provide an unexhaustible source of newness in the domain of research or in the academic world, where originality that helps to distinguish oneself is sometimes looked for? Was it perhaps a way to understand better some aspects of their central object of study — the general laws of human psychology, of linguistics, of neurology, etc.? Was it out of practical requirements — of educating normal children and of therapeutics of speech troubles? Was it the comfortableness (nevertheless apparent, when the investigation becomes scientific) of researchers who had daily under their eyes docile (again seemingly) and charming subjects?

All these, or part of them — if we refer to each researcher separately — were the real causes. And it is quite sure that all these causes persist and that they will be permanent impulses towards further studies in this field in Europe (or in any part of the world where specialists have leisure to embark upon research and where the most elementary tasks of didactic activity or of special investigations in the respective field have been exhausted and surpassed, or where requirements for suchlike studies do arise).

But the study becomes more and more detached from the affective motivations, it becomes ever less the concern of one parent and is taken over by the 'indifferent' specialist — anyhow, detached from the quality of being the parent of a certain 'case' — or even by teams of research workers. The own child is replaced by great collectivities of subjects under study, the observations become more systematic, the objectives more precise, and the interdisciplinary framing necessary. Applied linguistics, applied psychology, 'defectology', child psychiatry impose certain research themes, and theoretical linguistics or general psychology propose the verification of a number of modern theories or hypothesis through studying child language.

The European research in this domain began by being original and detached from the American linguistic conceptions of the various epochs; today one may notice (especially with younger research-workers) a taking over — not always creative — of theories or methodologic procedures from transformational or generative linguistics. To the extent in which these influences will be assimilated we may expect interesting research. It is likewise probable that a more precise psycholinguistic framing will be achieved, and it will thus be particularly useful (in this direction American research has now the advance as a movement of ensemble in child language research). Let us also expect the switch over from the exposition of results obtained from isolated cases, or from the investigation of details, to the great syntheses — but based on facts collected through the author's own investigations.

Perhaps the decade to come, or at least the 80's, will bring us also such a synthesis (at the same time, comparative for all

European languages), based both on the detailed research done until now in different European languages, and on the further investigations carried out with exact procedures and using an international and adequate transcription system, and done by big teams, on a large number of children, that would allow a statistical elaboration of data, the establishing of frequencies and the affirmation of the validity of the generalizations.

REFERENCES*

Bloch, Oscar,
1921 "Les premiers stades du langage de l'enfant", *JPsych.* 18, 693-712.
1923 "Langage d'action dans les premiers stades du langage de l'enfant", *JPsych.* 20:7, 670-72 (Discussions, 74).
1924 "La phrase dans le langage de l'enfant", *JPsych.* 21, 18-43.
Boons, J.-B., M. Coyaud, and C. Pichevin,
1966 "Acquisition du langage par les enfants de 20 mois à 4 ans environ: Grammaire et sémantique" (ronéo).
Borel-Maisonny, S.,
1960 Langage oral et écrit (Neuchâtel, Delachaux et Niestlé).
Brenstiern-Pfanhauser, Stefanja,
1930 "Rozwój mowy dziecka", *Prace Filolog* 15, 273-356 (Bibliography, 329-355).
Bühler, Charlotte,
1930 *Kindheit und Jugend*, 2nd ed. (Leipzig, Hirzel).
Bühler, Karl,
1926 "Les lois générales d'évolution dans le langage de l'enfant", *JPsych.* 23, 597-607.
1930 *Die geistige Entwicklung des Kindes*, 6nd ed. (Jena, Fischer), (Ch. 4: Die Entwicklung der Sprache, 220-247).
Cohen, Marcel,
1925 "Sur les langages successifs de l'enfant", *Mélanges linguistiques offerts à J. Vendryes* (Paris, Champion), 109-127.
1933 "Observations sur les dernières persistances du langage enfantin", *Psychologie du langage*, (1933), 390-399.
1951 Review of *Speech Development of a Bilingual Child* by: W. Leopold (1949), *Bull. Soc. Ling.* 47:135, 62-65.
1952 "Sur l'étude du langage enfantin", *Enfance* 3, 181-249.

* These References might be completed by our article, and its Bibliography: "Studiile europene asupra limbajului copilului (1920-1968)", *Studii şi cercetări lingvistice* 2, 3, 4, 5 (1969). (To appear in English in *Current Trends in Linguistics* IX, ed. by T. Sebeok [The Hague, Mouton]).

Cohen, Marcel, I. Lézine, F. Kocher, A. Brauner, L. Lentin and A. Tabouret-Keller,
1962 *Études sur le langage de l'enfant* (Paris, Éd. du Scarabée).
Čukovskij, Kornej,
1955 *Ot dvux do pjati* (Moskva, Sovetskij pisatel).
Decroly, O.,
[1933] *Comment l'enfant arrive à parler* [Bruxelles?] (I: *Cris-Babil-Compréhension, imitation, expression*; II: *Vocabulaire*; *Développement du langage*).
Delacroix, Henri,
1924a "L'activité linguistique de l'enfant", *JPsych.* 21, 4-17.
1924b *Le langage et la pensée* (Paris, Alcan).
1933 "Au seuil du langage", *Psychologie du langage*, 9-17.
1934 *L'enfant et le langage* (Paris, Alcan).
Francescato, Giuseppe,
1965 *Linguistica, psicologia e lo studio del linguaggio infantile* (Roma, Atti dell'Accademia Nazionale dei Lincei).
1970 *Il linguaggio infantile* (Torino, Einaudi).
Fronzaroli, Pelio,
1957 *Il linguaggio del bambino* (Bologna, Malipiero).
Grégoire, Antoine,
1933 "L'apprentissage de la parole pendant les deux premières années de l'enfance", *Psychologie du langage*, 375-89.
1937, 1947 *L'apprentissage du langage*, I: *Les deux premières années*, II: *La troisième année et les années suivantes* (Liège-Paris, [Fac. Philos. et Lettres, Liège-Droz]).
1948 "L'apprentissage du langage", *Lingua*, 1, 162-74; cf. Francescato, 1965: 403.
1949-1950 "La renaissance scientifique de la linguistique enfantine", *Lingua* 2, 355-398.
Guillaume, P.,
1927a "Les débuts de la phrase dans le langage de l'enfant", *JPsych.* 24, 1-25.
1927b "Le développement des éléments formels dans le langage de l'enfant", *JPsych.* 24, 203-229.
Gutzmann, H.,
[2nd ed. 1931; 1894] *Sprache und Sprachfehler des Kindes: Gesundheitslehre der Sprache* (Leipzig, Weber).
Gvozdev, A. N.,
1949 *Formirovanie u rebenka grammatičeskogo stroja russkogo jazyka* (Moskva, Izd. APN), 2 vol.
Jakobson, Roman,
[1939] "Les lois phoniques du langage enfantin et leur place dans la phonologie générale" [Communication présentée au Ve Congrès Int. des Linguistes, Bruxelles 1939], in N. S. Troubetzkoy, *Principes de phonologie* (Paris, Klincksieck, [1949] 1964), 367-379; and in R. Jakobson, 1962: 317-27.
[1941] *Kindersprache, Aphasie und allgemeine Lautgesetze*, in R. Jakobson, 1962: 328-401 (dated cf. p. 396).

1962 *Selected Writings*, I: *Phonological Studies* (The Hague, Mouton).

Jespersen, Otto,

[1922] 1959 *Language: Its Nature, Development and Origin*, 11th reimpression (London, Allen and Unwin) (Book II: *The Child*, 103-188).

1923 *Börnesprog: En bog for foraeldre* (Kobenhavn-Kristiania-London-Berlin, Gyldendaske boghandel).

Kaczmarek, L.,

1953 *Kształtowanie się mowy dziecka* (Poznań, Pozn. Tow. Przyj. Nauk) (Germ. abstr.: "Allgemeine Übersicht über die Gestaltung der Sprache des Kindes", 68-79).

Kaper, William,

1959 *Kindersprachforschung mit Hilfe des Kindes: Einige Erscheinungen der kindlichen Spracherwerbung erläutert im Lichte des vom Kinde gezeigten Interesse für Sprachliches* (Groningen, Wolters) (Fr. and Engl. abstr.).

1966 "Lässt sich beim noch nicht in die Schule gehenden Kind irgendein 'Gefühl' für die 'Leistung' des Genitivs und der Präpositionen feststellen?", *Lingua* 16:1, 40-56.

1967 "Einige Beobachtungen zur kindlichen Satzbildung", *Xth Int. Congr. Linguistics, Abstracts* (Bucharest, 1967), 170.

Kaverina, E. K.

1950 *O razvitii reči detej pervyx dvux let žizni* (Moskva, Medgiz).

Kenyeres, Elemer,

1927 "Les premiers mots de l'enfant et l'apparition des espèces de mots dans son langage", *Archives de Psych.* 20, 191-218.

1938 "Comment une petite Hongroise de sept ans apprend le français", *Archives de Psych.* 26, 321-366; cf. Leopold, 1952: 51.

Kostić, Đorđe, and Milja Stosić,

1963 *Akusticka struktura krika novorodenceta* (Beograd, Institut za eksperimentalnu fonetiku i patologiju govora).

Leopold, W. F.,

1952 *Bibliography of Child Language* (Evanston, Ill., Northwestern University Press).

1959 "Kindersprache", *Phonetica* 4:4, 191-214.

Lewis, M. M.,

1936 *Infant Speech: A Study of the Beginnings of Language* (London, Kegan Paul, and New York, Harcourt Brace).

Logopedia

1960-1969 (Zagadnienia kultury żywego słowa) (Lublin).

Luria, A. R.,

1928 *Reč i intellekt v razvitii rebenka* (Moskva), cf. Luria, 1959 (and: *Psychological Abstracts*, 1929: 515-517).

1955 *Rol' slova v formirovanii vremennyx svjazej v normal'nom i anormal'nom razvitii* (Moskva, Izd. APN).

1959 "Razvitie reči i formirovanie psixičeskix processov", *Psixologičeskaja Nauka v SSSR* (1959), 515-577.

Luria, A. R., and F. Ja. Judovič,
1956 *Reč i razvitie psixičeskix processov u rebenka*: *Eksperimental'noe issledovanie* (Moskva, Izd. APN).
Malmberg, Bertil,
1945 "Ett barn byter språk: Drag ur en fyraårig finsk flickas språkliga utveckling", *Nordisk tidskript för vetenskap* ..., 170-181.
Mikeš, Melanija,
1964 "Glasovni razvoj govora dvojezične dece (Novi Sad) (dissertation; stencilled, cf. M. Mikes).
1967 "Acquisition des catégories grammaticales dans le langage de l'enfant", *Enfance* 3-4, 289-299.
Nuffield Foundation,
[1964-1968] *Foreign Languages Teaching Material Project*: *Child Language-Survey, Reports and Occasional Papers* [Leeds].
O'Connor, N., and Beate Hermelin,
1963 *Speech and Thought in Severe Subnormality*: *An Experimental Study* (Oxford-London-New York-Paris, Pergamon Press).
Ohnesorg, Karel,
1948 *Fonetická studie o dětské řeči* [Étude phonétique sur le langage de l'enfant] (Praha, Filos. fakulta) (Fr. abstr., 175-179).
1955a "Růst slovní zásoby v dětské řeči" [L'accroissement du vocabulaire dans le langage de l'enfant], *Acta Univ. Carolinae* 1:1, 85-107 (Fı. abstr.).
1955b "Kapitola ze srovnávací fonetiky dětské řeči", *Sborník filos. fak. brněnské univ.* A 3, 95-101.
1959 *Druhá fonetická studie o dětské řeči* (Praha-Brno). (Fr. abstr. 149-159), Rus. abstr. 139-148).
1962 "Renesance studia dětské řeči", *Sborník filos. fak. brněnské univ.* B 9, 228-39 (Fr. and Russian abstr.).
1964a "Une contribution à la pédophonétique", Zeitschr. *Phonet.* 17:2-4, 265-267.
1964b "Étude sur le langage enfantin (une contribution à la pédolinguistique)", *Sborník filoz. fak. univ. Komenskéko Bratislava* 16, 19-30.
Oléron, P.,
1957 *Recherches sur le développement mental des sourds-muets*: *Contribution à l'étude du problème 'Langage et pensée'*. (Paris, C.N.R.S.).
Osgood, Ch., and Th. Sebeok (eds.)
1954 *Psycholinguistics*, 2nd ed. (Bloomington, Indiana University Press, and Baltimore, Waverley Press).
Pavlovitch, Milivoïe,
1920 *Le langage enfantin*: *Acquisition du serbe et du français par un enfant serbe* (Paris, Champion).
Piaget, Jean,
1923 *Le langage et la pensée chez l'enfant*: *Études sur la logique de l'enfant*, Préface de Ed. Claparède (Neuchâtel-Paris, Delachaux et Niestlé), 2nd ed., 1930.
[1924] *Le jugement et le raisonnement chez l'enfant* (5-81; Ch. I, Grammaire et logique: L'emploi des conjonctions de connexion causale et logi-

que et des conjonctions de discordance chez l'enfant de 3 à 9 ans) (Neuchâtel-Paris, Delachaux et Niestlé).
Psychologie du langage,
1933 (Paris, Alcan).
Raffler-Engel, Walburga von,
1964 *Il prelinguaggio infantile* (Brescia, Paideia).
Ronjat, Jules,
1913 *Le développement du langage observé chez un enfant bilingue* (Paris, Champion).
Rūķe-Draviņa, Velta,
1963 *Zur Sprachentwicklung bei Kleinkindern*, 1: *Syntax* (Lund, [Slaviska Inst. vid Lunds Universitet]).
1967 *Mehrsprachigkeit im Vorschulalter* ([Lund], Gleerup).
Rybnikov, N. A.,
1926 *Jazyk rebenka* (Moskva, Gosizdat); cf. Brenstiern-Pfanhauser, 1930: 347.
Rybnikov, N. A. (red.)
1927 *Detskaja reč: Sbornik statej* (Moskva); cf. Brenstiern-Pfanhauser, 1930: 348.
Slama-Cazacu, T.,
1957 *Relaţiile dintre gîndire şi limbaj în ontogeneză* (3-7 ani) [Relations entre la pensée et le langage en ontogenèse (3 à 7 ans)]. (Bucureşti, Ed. Acad. R.P.R.) (Russian and Fr. abstr.).
1960 "Aspecte ale relaţiilor dintre gîndire şi limbaj în însuşirea structurii gramaticale de către copii ante-preşcolari (2-3 ani)", *Rev. Psihol.* 2, 43-65 (Russian and Fr. abstr.: Aspects des relations entre la pensée et le langage dans l'assimilation de la structure grammaticale chez le petit enfant [2-3 ans]).
1961 *Dialogul la copii* (Bucureşti, Ed. Acad. R.P.R.) (Russian and Fr. abstr.: Le dialogue chez les enfants) (Czech ed.: *Dialog u dětí.* [Praha, 1966]. Preface by V. Chmelař and K. Ohnesorg).
1962a "The Oblique Cases in the Evolution of Child Language", *Rev.roum. ling.* 1, 71-90 (and *SCL* 2 [1961], 215-229).
1962b "Particularităti ale însuşirii structurii gramaticale de către copii, între 2 si 3 ani", *Culegere de studii de psihologie* IV (Bucureşti, Ed. Acad. R.P.R.), 119-152.
1965 "La dynamique des relations entre la pensée et le langage", *Rev. roum. sci. soc.* — Psychologie 2, 167-180.
1966a "Le dialogue chez les petits enfants: Sa signification et quelques-unes de ses particularités", *Bull. Psych.* 19 (247):8-12, 688-697.
1966b "Quelques remarques théoriques et méthodologiques sur le problème de la forme et de la substance dans la genèse du système phonématique", *Cahiers de ling. théor. et appl.* 3, 171-179.
1967 "Sur la formation du système phonématique chez l'enfant", *To Honor Roman Jakobson* III, (The Hague, Mouton), 1857-1865.
1968 *Introducere în psiholingvistică* (Bucureşti, Ed. ştiinţifică) (Engl. abstr.: Introduction to Psycholinguistics); in print, in Italian—Bologna,

Pátron —, and in English — Bucharest - The Hague, Ed. ştiinţifică — Mouton.
Stern, Clara, and Stern, William,
1928 *Die Kindersprache: Eine psychologische sprachtheoretische Untersuchung*, 4th ed. [neubearbeitete Auflage] (1st ed. 1907; 3rd ed. 1922) (Leipzig, Barth).
Stosic, Milja,
1964 *Razvitak dečjeg recnika* 1 (Beograd, Institut za eksperimentalnu fonetiku i patologiju govora).
Tabouret-Keller, Andrée,
1963 "L'acquisition du langage parlé chez un petit enfant en milieu bilingue", *Problèmes de psycholinguistique* (Paris, PUF), 205-219.
Tervoort, Bernard Th.,
1953 *Structurele analyse van visueel taalgebruik binnen een groep dove kinderen* [Structural Analysis of Visual Language Use with a Group of Deaf Children] (Amsterdam, Noord-Hollandsche Uitgeversmij), 2 vols.
1961 "Esoteric Symbolism in the Communication Behavior of Young Deaf Children", *Amer. Annals Deafs*, 106:5. 436-480.
Van Ginneken, J.,
1917 *De roman van een kleuter* (Nijmegen, Malmberg).
1924 *De ontdekkingen van den kleuter* (Utrecht-Nijmegen, Dekker, Vegt and Leeuwen).
Vasič, Smiljka,
1965 "Govor dece u obdanistu" ("Nadezda Purić" u Beogradu, Beograd Inst. za eksper. fonet. i patolog. govora) (stencilled).
Vendryes, J.,
1953 Review of M. Cohen et al., *Sur l'étude du langage enfantin* (1952), *Bull. Soc. Ling.* 49:139, 25-27.
Vygotskij, L. S.,
1956 *Izbrannye psixologičeskie issledovanija: Myšlenie i reč, Problemy psixologičeskogo razvitija rebenka* (Moskva, Izd. APN). (Introduction by A. N. Leont'jev and A. R. Luria 4-36).
[1962] *Thought and Language*, transl. by E. Hanfmann and G. Vakar (Cambridge, Mass., M.I.T. Press).
Siblé, Eugène,
1954 Review of M. Cohen *Sur l'étude du langage enfantin* (1952), *Cahiers F. de Saussure* 12, 75-77.

TOMOANALYSE ET OSCILLOGRAPHIE

SUZANNE BOREL-MAISONNY
(Paris)

POSSIBILITÉS D'APPLICATION AUX TROUBLES DE LA PAROLE

Il y a de nombreuses années quand, aux enregistrements sur cylindre passé au noir de fumée, furent substitués peu à peu les enregistrements oscillographiques sur film, ce fut un progrès évident. A la place d'un système entaché d'inertie, l'assemblage microphone-tube cathodique-caméra fournit de grandes et belles images. Il suffisait que le film se déplaçât à vitesse convenable pour avoir une représentation fidèle de la parole. Personnellement, je fus immédiatement séduite et l'œuvre qui me donna cette première joie fut celle de Gemelli, *Analyse électro-acoustique du language*, travail complété ultérieurement en 1936. Dès 1936, je pus avoir une installation qui donna lieu à des publications dans la *Revue Scientifique*, dans des journaux médicaux et dans *Le Français Moderne*. Voici quelles sont les caractéristiques de ma propre installation: un microphone capte les bruits de la parole qui, sur l'écran d'un tube cathodique, se manifestent par des mouvements du spot. Un film se déplace à vitesse constante devant ce point, il en résulte un tracé qui donne en quelque sorte une représentation visuelle de la parole.
Le film est de 34 mm, les bobines de la caméra peuvent en recevoir 10 mètres.

Un repère de temps lumineux, produit par un disque à 100 fentes qui tourne synchroniquement avec le moteur de la caméra, s'inscrit sur le film en petits traits équidistants. L'intervalle entre deux traits représente donc une durée de 1/100 de seconde. L'entraînement des bobines est d'autant plus rapide que le patin de l'embrayage est plus près de l'extérieur du disque ce qui, avec les diverses possibilités du moteur, se traduit pratiquement par des vitesses de déplacement du film allant de 5 millimètres à 50 millimètres au centième de seconde.

On voit donc ce qui s'est passé acoustiquement pendant une durée de temps exprimable en centièmes de seconde, dixièmes de seconde et secondes sur un tracé dont la longueur se mesure en millimètres, centimètres et mètres.

La fidélité d'un tel appareillage est fonction de la qualité du micro, du système optique et de la précision mécanique de l'ensemble moteur dont les défaillances possibles sont controlables par les variations du repère de temps.

De ces oscillogrammes, nous avons une collection considérable ... de plusieurs milliers de mètres. Constituée dans des conditions différentes pour répondre aux diverses questions que nous nous sommes posées, elle comporte des documents donnant des renseignements d'ordre phonétique, d'ordre linguistique et d'ordre acoustique.

Nous dirons — sans paradoxe — que le malheur de l'oscillogramme est de pouvoir donner une image fidèle de la parole, quant à la durée de ses éléments, à leur intensité et à leur variabilité en fonction du temps, si bien que les moindres fluctuations sont inscrites dans le tracé qu'il y a lieu de lire. Il y faut une certaine expérience et, comme pour toute instrumentation, en bien connaître les possibilités et les limites: les appareils sont d'ordinaire moins en défaut que ce que nous prétendons indûment leur faire dire.

Le principal inconvénient de l'oscillogramme est donc la longueur de l'image. Un film se déroulant à la grande vitesse de 50 mm au centième de seconde pour 10 m de film contient 9 m de tracé valable car il faut OTER LE TEMPS D'ÉTABLISSEMENT D'UNE VITESSE RÉGULIÈRE et le TEMPS DE RÉACTION du locuteur au micro. A cette grande vitesse, le texte qu'il sera possible d'enregistrer aura donc une durée de $\frac{1 \text{ cs} \times 9000}{50}$ soit 180 centièmes de seconde — MOINS DE 2 SECONDES — mais en utilisant la vitesse la plus réduite, le temps de parole enregistrable (la longueur du film inutilisable étant réduite) sera donc de $\frac{1 \text{ cs} \times 9500}{5} = 1900$ centièmes de seconde = 19 SECONDES. En fait, bien qu'il faille, pour avoir un beau tracé lisible adapter la vitesse à la hauteur et aux caractéristiques de la voix (des voix aiguës demandant un déroulement plus rapide que les voix graves et les voix au tracé riche plus d'étalement que les voix à l'expression graphique simple) les vitesses de déroulement le plus couramment utilisées vont de 1 cm à 3 cm au centième de seconde permettant, pour un film de 10 m, des durées de parole de 3 à 9 secondes 1/2.

LE CHOIX DE LA VITESSE dépend surtout, évidemment de ce qu'on cherche à voir. S'il s'agit de renseignements sur LE RYTHME, LES DURÉES SYLLABIQUES RESPECTIVES, LES VARIATONS D'AMPLITUDE, ce sont les vitesses réduites qu'il faudra utiliser; si l'on cherche des renseignements sur LES VARIATIONS DE TIMBRE, D'ATTAQUE OU D'EXTINCTION DES SONS, sur LES TRANSITIONS PHONÉTIQUES, sur L'INTÉRACTION DES PHONÈMES, on utilisera un déroulement plus rapide; les grandes vitesses seront réservées aux fins détails relatifs à la fréquence et les vitesses très réduites à la mise en valeur des caractéristiques de la chaîne parlée.

On se rappellera toutefois que le temps se traduisant en espace, si cet espace est trop condensé, le tracé est ininterprétable mais s'il est trop étalé nous n'avons plus de perception de l'ensemble. J'ai donné ici les caractéristiques de ma propre installation. J'en ai tiré le plus clair de mes renseignements quant à la phonétique de la parole, renseignements qui, joints aux investigations endoscopiques et radiographiques — contrôlés ultérieurement par la tomoanalyse — constituent une base vraiment solide quant à la connaissance de la structure de la parole. En revanche, l'oscillographie est peu adaptée à la connaissance acoustique des phonèmes, bien que les courbes interprétées avec prudence, fournissent sur ce point aussi une abondante moisson. Et bien entendu, les problèmes de perception n'y sont qu'effleurés, mais beaucoup sont quand même posés par le simple aspect des tracés.

INTERPRÉTATION DES TRACÉS

Quels sont donc les renseignements qui dans les conditions ci-dessus indiquées, peuvent être obtenus quant à la connaissance de la parole?

On se rappellera qu'un tracé n'est interprétable que s'il est accompagné d'un repère de temps, s'il est suffisamment lisible et pris à une vitesse permettant la compréhension de ce qu'on cherche. La fig. 1 met en évidence des variations de forme en fonction de la vitesse de déroulement.

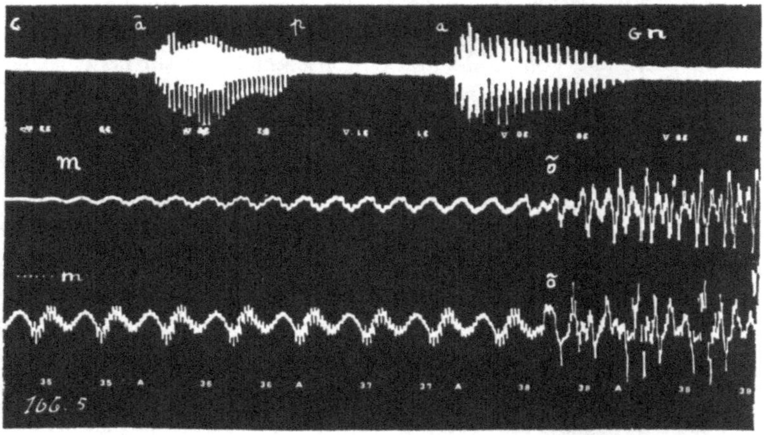

Fig. 1. Ces trois tracés sont pris à des vitesses différentes. Celui du haut de la figure représente le mot *campagne*. Le repère de temps correspond à un écoulement du film égal à 3 millimètres au centième de seconde. Le tracé du milieu est celui d'une voix de femme et le tracé du bas une voix d'homme pris à la même vitesse de 20 mm au centième de seconde. Le repère de temps a été coupé pour grouper les images. On remarquera que l'*m* de *mon* en voix masculine est plus long que celui en voix de femme. Texte prononcé: *mon chien blanc*.

Le bruit de fond doit absolument être évité sous peine qu'on ne puisse apprécier ce qui se passe dans les nombreux moments de silence de la parole, silences PAUSAUX — intentionnels ou réels et dans les temps de tenue des occlusives sourdes, notamment.

Nous avons publié à plusieurs reprises des fragments d'oscillogrammes; un seul en entier, le N° 136 ns de notre collection et cela pour des raisons faciles à saisir: le tracé est long. En extraire les détails défigure l'ensemble. On ne comprend plus ..., alors que le déroulement de la totalité du texte parlé en offre une reconstitution visuelle saisissante.

L'UNITÉ DE PAROLE: la syllabe. Son expression centrale est la voyelle. La parole se présente comme un ruban sonore qui paraît continu, surgissant du silence ou se situant entre des temps de repos, que sont les pauses et les stops entre les rhèses.[1]

En fait, de tout ceci nous avons une impression globale approximative très suffisante pour diriger notre discours ou maintenir l'attention que nous portons à celui d'autrui, mais il est impossible de chiffrer ces durées dont nous avons une idée parfois assez erronée. Personne, s'il ne l'a contrôlé par des moyens appropriés, n'a une idée de la durée totale de ses discours et encore moins des mots phonétiques ou phoniques, des syllabes et des phonèmes qui le composent. Pourtant, nous ne sommes pas insensibles à un certain tempo de ces durées et sommes également gênés — si, en dehors de toute manifestation pathologique —, un débit est plus rapide ou plus lent qu'il n'est en usage dans le parler des personnes qui nous entourent.

Quelles sont donc ces durées qui s'imposent à notre perception tandis que d'autres paraissent absolument lui échapper?

A distance, on entend des successions de coulées verbales qui paraissent continues. 'On parle' mais ce qui est dit nous échappe.

[1] On appelle de ce nom les ensembles parlés constitués par la signification attachée à nos paroles et qui constituent en quelque sorte une analyse de la structure sur le plan de l'expression; ex.:

Le chien vient de se sauver ... essayez vite de le rattraper. Où sont mes gants; je crains de les avoir perdus.

Il n'y a plus de mots séparés, les groupements varient avec l'intention expressive.

En se rapprochant on entend des éléments sonores vocaliques qui se différencient: la parole prend un aspect organisé; de plus près, des 'consonnes' se précisent: un sens se dégage et dans une même rhèse ou fragment signifiant de parole, on distingue des variations d'intensité grâce aux unités syllabiques. Elles sont plus ou moins sensibles, suivant la nature des phonèmes qu'elles enclosent.

Dans *il a eu peur*, l'amplitude va du minimum *p* au maximum *œ̀*; *cet homme est parti* [sètòmèparti] offre des limites syllabiques qui s'imposent également sans peine, créant les répartitions *i, la, u, pœ̀r*; *sè, tò, mè, par, ti* alors que, dans des phrases telles que *qu'est-ce qu'il y a eu*; *donne-moi de l'huile*, elle est plus discutable car on pourrait à la rigueur découper ces deux phrases en: *kès kil ya u*; *dòn mwad lwil* ou *kè ski lya u*; *don mwa dlwil.*

De toute manière, la voyelle est le centre de la syllabe à moins qu'elle n'en soit le tout (ex.: *eu* (*u*) part. passé du v. *avoir*).

Quant à la DURÉE DES PHONÈMES qui s'inscrivent dans cette chaîne parlée, elle est de l'ordre de 10 à 12 centièmes de seconde en moyenne, mais avec des variations très importantes s'inscrivant entre les extrêmes allant de 2-3 centièmes de seconde pour les plus brefs, jusqu'à 30 et 40 centièmes de seconde pour les plus longs, ceci dans le discours normal où ne figurent ni les *œ* ..., remplacés parfois par des chevilles, (*n'est-ce pas, vous comprenez, et alors, dans un certain ordre d'idées*), ni les accélérations des tachylaliques.

De ces variations et excès de vitesse, nous donnons des ex. f. 2. Si les DOCUMENTS SONORES recueillis au cours des conférences étaient explorés sur ce point, on se rendrait compte de ces irrégularités; ELLES GÊNENT RAREMENT LA COMPRÉHENSION et TEL DISCOURS SUIVI SANS PEINE À L'ÉCOUTE DIRECTE sans qu'on ait même conscience que parfois il boitait, S'AVÈRE, TRANSCRIT D'UNE BANDE MAGNÉTIQUE, constellé de parasites et, par instants, ininterprétable mot à mot.

LES PHONÈMES posent un autre problème: celui de leur perception. Il arrive qu'ils soient réduits à 2 ou 3 centièmes de seconde mais ce ne sont pas n'importe lesquels: *l, r, n,* sont les plus atteints

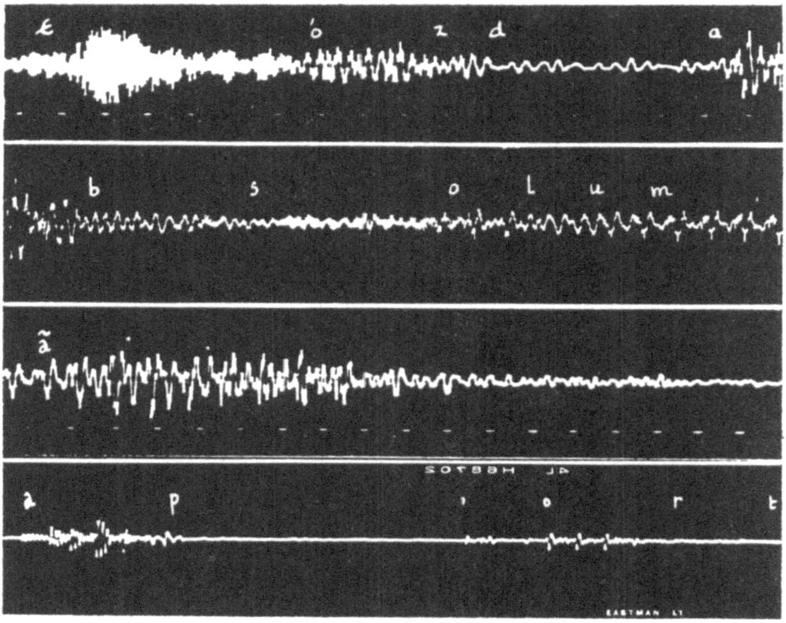

Fig. 2. Irrégularités du débit chez un homme de 30 ans racontant une histoire animée; fragment de parole représenté: "*quelque* chose d'absolument *extraordinaire*". Le repère de temps permet de constater que dans ce passage il y a des phonèmes particulièrement brefs: *z a o l u*. On constate aussi une tendance nette à faire durer les consonnes plus que les voyelles. La consonne *l* a duré environ 2 centièmes de seconde.

en certaines positions: *il n'y a pas de pain* se rapprochant tellement de *i nya pad pè* ou même [l], [n] *y a pas de pain* et ce type de phrase a pu engendrer: *yapádpe*; de même pour: *elles ne sont plus là* [è (l) nœ sõ plu la] souvent émis: *ènsõplula*, et même ... *èsõpula*. C'est à peine si ces omissions sont perçues, tellement le sens s'impose à nous, mais cela explique chez l'enfant la reproduction lacunaire qui se maintient chez les sujets ayant du retard de parole; exemple: *la pot è fèmé* (*la porte est fermée*) *i sé pa di* (*r*) *sa* (*il ne sait pas dire ça*), *le ti na* (*r*) *b* (*le p'tit arbre*), *letitarb* ... etc.

Or, si ces mêmes éléments réduits manquent accidentellement dans le discours d'un sujet en âge de bien parler, nous nous apercevons immédiatement de ce relâchement phonétique. Des éléments de 2 centièmes de seconde seraient-ils donc perçus pour que leur absence réelle soit aussitôt sentie? Nous pensons que chaque phonème est préparé par le précédent et s'amalgame en quelque sorte avec le suivant, c'est donc cet enchaînement que nous percevons. Les oscillogrammes, à cet égard, sont des documents irrécusables, mais ils prouvent ainsi que les préparatifs du phonème ne suffisent pas à son identification. Si, dans le discours, la prévisibilité de ce qui suit nous fait 'entendre' ce qui n'existe pas, dans des syllabes sans signification, il n'en est pas de même et nous ne courons pas le risque d'entendre *irbe irme irne*, si le locuteur a prononcé: *ibe*, *ipe*, *ime*, *ine* ... et nous n'identifions *b p m n* qu'après leur finition.

Ce que les oscillogrammes nous montrent, quant à la présence réelle de certains phonèmes et à leur durée, est confirmé par une écoute en TOMOANALYSE, c'est-à-dire dans un texte découpé en tranches sonores, partiellement superposables, de telle sorte qu'un fragment de présentation est réentendu avec le début du suivant et ainsi de suite, réalisant ainsi une reproduction ralentie et continue de tout le texte parlé.

LES ALTÉRATIONS DES PHONÈMES EN CONTACT

La simple audition, dans les conditions d'écoute normale, nous fait mauvais juges de ce que nous entendons, et un enregistrement oscillographique montre qu'il se passe quelque chose d'inentendu, parfois très facile à interpréter, mais pas toujours. Il est exceptionnel que d'autres techniques ne confirment pas la réalité du phonème, dont la présence est révélée par l'oscillogramme; le découpage analytique sonore est un excellent moyen de contrôle d'états transitoires dont les étapes constatées sont alors entendues et qu'on a ainsi le loisir d'écouter autant de fois qu'il faut et d'analyser ensuite.

Sur quels faits avons-nous cherché des précisions? En voici des échantillons:

(1) Sur les consonnes en JOUXTURE ou en SÈVREMENT dirions-nous, en employant le vocabulaire de Pichon. En d'autres termes, y a-t-il le même traitement phonétique — donc articulatoire — dans: *Jacqueline*, *Claire*, et *claque la porte* où *kl* se trouve associé à une voyelle. Le cours *Hattemer* présente-t-il le même traitement de *tm* que *prête-moi* ou l'*ethmoïde*; ou encore *les Schleus* et *il pêche la truite*, sont-ils identiques dans l'articulation du groupe *chl*,? ... et ainsi de suite ...

(2) Qu'est exactement la séquence consonne sourde — consonne sonore dans notre parler coutumier — non teinté d'accent méridional, bien entendu — dans des mots phonétiques où sont inclus des assemblages tels que: *le médecin* (*ds*), *elle brode son mouchoir* (*ds*), une *robe teinte* (*bt*), *l'absence* (*bs*), *bouche cousue* (*chk*), *il passe tard* (*st*), *une rose thé* (*zt*), *tu ne déranges pas* (*jp*), *une touffe verte* (*fv*), *une cave fraîche* (*vf*), etc., etc.

(3) Comment s'opère le glissement de *ch*——→*s* (*il couche son chien*) ou de *s*——→*ch* (*il passe chez moi*) ou de *z*——→*j* (*un vase jaune*) ou de *j*——→*z* (*les songes africains*) ...

(4) Comment se traitent, sur le plan de l'articulation, les groupes *gl bl dl* ou *gr br dr* (*les ongles*, *une glace bleutée*, *elle borde le lit*, *de gros bras dressés de joie*, etc.).

(5) Que se passe-t-il dans une gémination du type: *il jette tout* [èttu], *une robe bleue* [obbloe], *il tape partout* [appa], *il laisse sa veste* [èssa], *un prof furieux* [offu], *une grande dame* [ãdda].

(6) Que se produit-il lorsqu'une voyelle nasale s'installe à partir du silence: *entre* [ã]; ou est située près d'une ou deux voyelles orales: *du thon à l'huile* [õa], *va en haut* [ãó]; en position finale: *il ment* [ã] ou lorsqu'elle est en contact avec des consonnes à caractéristique orale ou nécessitant une communication bucco-nasale: *enfin* [ãfẽ], *un pan de muraille* [ãdœm ou ãdm], *le pont tremble* [õtrã], *ventre à terre* [ãtra], *maman m'aime* [ãmè], *une annonce* [anõ], *un nid nouveau* [ãninuvó]. Ces deux derniers groupements, riches en nasales, permettent-ils une oralisation convenable des phonèmes théoriquement exempts de nasalité? Et ainsi de suite.

(7) Comment finissent normalement les phrases en fonction

de l'intonation, de la place dans le discours, du rythme et du contenu phonétique? Ce qui EST véritablement et ce qu'on CROIT EXISTER ne se superposent pas toujours.

(8) La durée syllabique est difficile à apprécier, plusieurs facteurs difficiles à dissocier se présentant ensemble: accent de longueur et d'intensité notamment, par exemple: L'INTENSITÉ RÉELLE n'est pas L'INTENSITÉ SUBJECTIVE.

(9) Quelles sont et doivent être les manières d'attaquer les voyelles et les consonnes? Que peut-on, en ce domaine, considérer comme normal?

(10) Que peut-on considérer comme étant de règle dans les rapports d'amplitude des voyelles et des consonnes et dans la distribution des accents d'intensité? En d'autres termes, une étude des variations accentuelles en fonction d'un type de parler est possible.

(11) Comment peut-on mesurer la vitesse de déroulement du discours et quelle est la distribution des pauses ou ponctuation orale? A partir de quels faits peut-on considérer qu'on sort du normal?

(12) L'ensemble des voyelles et des phonèmes sonores, dans un discours est toujours plus long que celui des phonèmes sans bruit laryngé audible. Mais quelques consonnes seulement étant toujours sourdes (*p t k*, *ch s f*, toujours, et *l y r*, en certaines positions), comment peut-on avoir l'impression d'un son continu? Enfin, les oscillogrammes montrent que l'extinction des bruits laryngés se fait graduellement pendant la tenue des consonnes sourdes, le silence total n'est obtenu que pendant une partie de la tenue des consonnes sourdes.

(13) Les voyelles ont un spectre inégalement stable. Les voyelles à l'articulation vigoureuse sont plus constantes dans leur forme, par exemple: les voyelles longues et les nasales ont un timbre plus fluctuant.

Les oscillogrammes ont donc servi à l'exploration systématique des PHONÈMES dans toutes les positions, de PHRASES COURTES, de PETITS DIALOGUES, de TEXTES que, parfois, nous avons répété à de multiples exemplaires en changeant de locuteurs.

Voici, à titre d'exemple, quelques textes, non choisis au hasard, mais pour leur intérêt phonétique, rythmique, etc.

— *Un chapeau doublé bleu avec grosses clématites, deux cents francs* (R. Benjamin, *Justices de Paix*).
— *Le vent a provoqué un raz de marée, n'est-ce pas?* (Veit)
— Un long texte: *Les "Anes de l'autre monde"* (P. Arène).
— "*Mais, bâtard de sort, Monsieur le Curé, comment ne voulez-vous pas que l'on grogne, quand ce sont toujours les mêmes qui travaillent et toujours les mêmes qui ne font rien ... et quand tout le temps, sur l'échine on sent l'écorchure du bât!*"

— *Le brou de noix est brun.*
— *Un i typique et difficile, c'est le minimum ...*

— *Le vent qui devient froid et la mer qui déferle*
Tout me dit que l'hiver est arrivé pour moi
Et qu'il est temps enfin, ramassant mes olives
D'en offrir l'huile vierge
A l'autel du Bon Dieu. (Mistral, *Les Olivades*)

— *Où est l'verre incassable?*
Dans l'buffet, avec les autres!
Ah! par exemple, je ne le vois pas ...

— *C'est une affaire intéressante, qu'en pensez-vous?*
Il faut la faire sans aucun regret.

— *Où as-tu mis le pâté de canard?*
— *Où est la pâtée des canards?* etc., etc.

Sur tous ces sujets, des informations précises peuvent être apportées par des tracés correctement exécutés.

Ces recherches ne traduisent-elles qu'un intérêt de curiosité et sont-elles seulement aptes à étoffer des discussions de phonéticiens? Il serait déjà légitime de les avoir poursuivies pour connaître plus intimement — en dehors de toute contrainte sérieuse — c'en est une si petite que de parler devant un micro! — comment s'organise l'expression orale du discours et comment, à travers variantes et lacunes, les éléments constants et stables finissent par être électivement toujours perçus.

Mais si l'on doit examiner des cas présumés pathologiques, il n'y a pas d'autre ressource que la comparaison avec les états considérés comme normaux, le mécanisme des uns et des autres pouvant être ainsi apprécié correctement.

On n'oubliera pas, toutefois, que la prise d'un document sonore valable est beaucoup plus difficile, dans certains cas pathologiques, que chez un sujet réagissant normalement et que la lecture en est incomparablement plus difficile, certains tracés de parole pathologique seraient d'inexplicables rébus sans l'audition, répétée à satiété, du document sonore.

CRITIQUE DU SYSTÈME

Deux défauts à ces graphiques:

(1) ILS REPRÉSENTENT LE SON ENTENDU PAR L'OREILLE, MAIS NE LE SITUENT PAS et n'apportent donc, qu'un RENSEIGNEMENT GLOBAL complété ultérieurement, par l'enregistrement au niveau du larynx, de quoi sont issus les glottogrammes. L'adjonction, par certains expérimentateurs, de petits micros à l'orifice narinaire ou intrabuccaux, malgré des complications supplémentaires sont utiles pour la vérification de quelques défauts, mais ayant l'inconvénient d'encombrer, ils annulent un des principaux avantages de l'oscillogramme qui respecte la liberté du locuteur: parler devant le micro est une infime entrave!

(2) L'autre défaut est plus sérieux, c'est l'ABSENCE D'ANALYSE SPECTRALE, dans une certaine mesure du moins. Si des renseignements peuvent être déduits du simple examen de la fig. 3 et calculés, d'autres informations impliqueraient une analyse fréquentielle de surcroît. Si la figure a permis de compter le nombre des vibrations aiguës par rapport au fondamental, lui-même situé en fréquence par rapport au repère du temps, il n'en est pas moins vrai que l'application d'une formule de Fourier à ces courbes est nécessaire pour savoir ce qui se passe, et encore certaines conditions doivent être remplies.

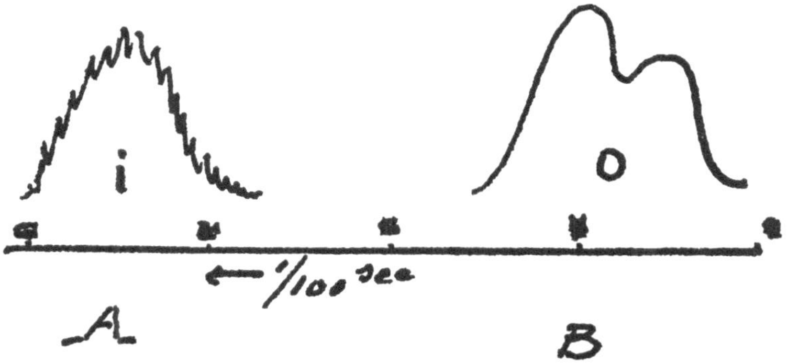

Fig. 3. Elle représente une période d'*i* et une période d'*o* à la même vitesse de déroulement du film; la voix est grave: environ 80 Hz dans ce fragment. Or s'il est facile de calculer la hauteur des vibrations aiguës de l'*i*, la structure de la période isolée d'*o* ne peut être appréciée que par une analyse de fréquence.

Cette lacune est à l'origine de la vogue pour les sonagrammes, qui donnent spatialement une analyse immédiate des sons. Les premiers essais, dénommés *Visible Speech* témoignent de cette qualité essentielle. Après de premières analyses spectrales très imparfaites, de successives améliorations ont porté sur l'INTENSITÉ — désormais objet d'une courbe spéciale et non plus seulement déduite de la saturation de l'image, — SUR LA FRÉQUENCE dont le champ primitif a été étendu de 3500 Hz à des hauteurs plus élevées qu'il n'est nécessaire pour la voix humaine et des artifices tels que la possibilité d'étaler plus ou moins l'image en fréquences et dans le temps ont été d'importants perfectionnements. La servitude principale demeure le temps de 2 sec. 4 comme limite de la durée de l'objet sonore. Les images ont un caractère flou et, suffisantes pour une vue globale et analytique, elle appellent cependant le désir d'autres précisions que vont donner entre autres les analyses de fréquence sur 3 paramètres: INTENSITÉ, HAUTEUR et VOLUME SONORE développé dans le temps.

EXPLORATION RADIOLOGIQUE COMPLÉMENTAIRE

L'image oscillographique implique une transcription phonème

à phonème des variations du tracé. Nous avons gardé pour la langue française l'alphabet Rousselot auquel nous avons adjoint quelques signes diacritiques et, au besoin, un commentaire, mais savoir attribuer des phonèmes aux variations d'un tracé ne donne pas de précision sur les conditions articulatoires de ces phonèmes. L'exploration radiologique nous en donne la clé. Les radioscopies en position statique et le film radiographique pour la chaîne parlée nous renseignent sur les attitudes et mouvements essentiels et ceux qui sont accessoires. Ces derniers sont variables, les premiers sont constants pour un parler déterminé. L'influence de la conformation interne des sujets est indéniable: elle n'est qu'une adaptation aux conditions nécessaires d'émission pour un timbre vocalique donné. Ce qui est déterminant, ce sont les rapports des cavités. L'aperture buccale est secondaire: je puis émettre toutes les voyelles avec une ouverture de bouche invariante en gardant, pour toutes ces voyelles, un écartement intermaxillaire constant, soit très petit (Fig. 4 I) soit très grand (Fig. 4 II) et en tenant les

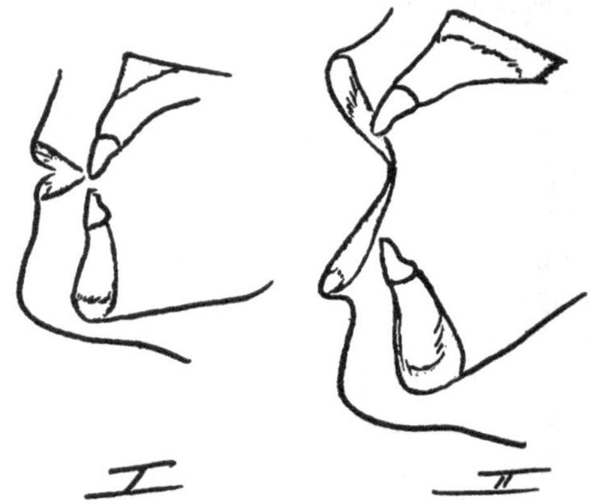

Fig. 4. I et II. En gardant pour la mâchoire et les lèvres une immobilité absolue, on constate que la position I permet l'articulation de toutes les voyelles qui demeurent parfaitement reconnaissables. Dans la position II, toutes les voyelles sont distinctes sauf *on ou u* pour lesquelles sans qu'il soit nécessaire d'ouvrir moins la bouche, une minime labialisation suffit.

lèvres immobiles dans l'attitude indiquée par la figure correspondante. La différenciation des timbres est parfaite et aisée dans l'attitude I. Elle est encore possible dans l'attitude II. Mais l'absence de labialisation est une grande gêne pour la production d'un *u* et d'un *ou* vraiment caractérisés; en outre *ou* ne sera vraiment *ou* qu'à partir d'une cavité suslaryngée importante et d'une zone de rétrécissement au-dessus; de même, *i* ne sera *i* que si une grande cavité postérieure est laissée libre par l'avancement de la langue qui doit en même temps monter vers le palais; *a* postérieur [á] est la voyelle qui entraîne le plus l'arrière de la langue vers la paroi postérieure du pharynx, etc.

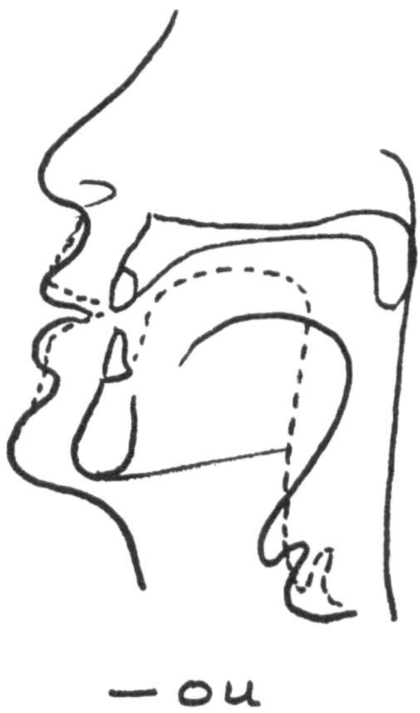

Fig. 5. Comparaison de l'attitude linguale pour les deux voyelles *u* et *ou*.

Sur le plan pédagogique, nous avons, à partir de ces explorations, fait un catalogue de mouvements comparés relativement à l'articulation vue aux rayons X, et même, pour les petits enfants, des portraits pour aider à la réalisation de la mimique la plus favorable à l'émission des phonèmes.

Il n'y a aucun inconvénient à introduire, dans les tracés représentatifs de la parole, une traduction vibratoire uniforme de toutes les voyelles. Leur différenciation s'apprend par la connaissance des mouvements et des attitudes.

Les tracés artificiels représentatifs de la parole

Ils s'inspirent à la fois des oscillogrammes et des tracés du type Rousselot sur noir de fumée pour donner, en raccourci, une représentation de l'effet à obtenir afin de parler juste; des passages rectilignes identiques indiquent les occlusives (ce n'est pas là qu'on apprend à les différencier).

De la connaissance des oscillogrammes et des kymogrammes, nous avons tiré une représentation abrégée des tracés de la parole, dont nous allons donner quelques exemples; ils permettent d'accompagner l'écriture phonétique d'un commentaire permanent des mouvements à accomplir et des résultats à obtenir.

TOMOANALYSE

Les difficultés d'interprétation des oscillogrammes sont considérablement diminuées par la méthode de la tomoanalyse. Il s'agit, en effet, d'un découpage sonore grâce auquel, au fur et à mesure qu'on parcourt le texte, il est possible d'en avoir, auditivement et sans déformation, une image ne laissant dans l'ombre aucun détail. Ce résultat est obtenu par un artifice: le fragment écouté a une certaine étendue et les fragments successifs se recouvrent partiellement.

Ces présentations doivent principalement tenir compte de deux facteurs:

(a) Il faut que la durée du fragment soit suffisante pour que l'oreille ait le temps de percevoir ce qu'il y a dedans, mais il ne doit pas être trop long sous peine de ne pas mettre en évidence le détail à observer.

(b) Ces fragments sonores doivent se succéder à des intervalles tels que le souvenir en soit encore net quand le fragment suivant

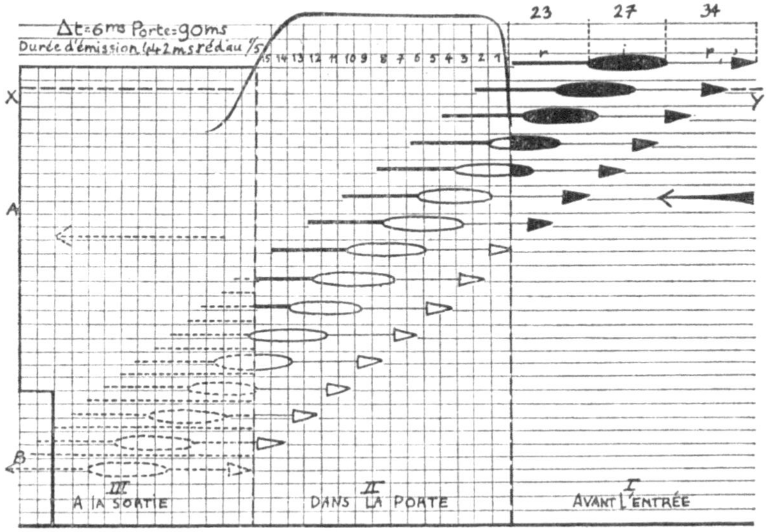

Fig. 6. Passage de la syllabe *rip* dans la porte électronique. Les durées respectives des phonèmes pour cette syllabe, mesurées en nombre de présentation, sont de 23 (*r*), 27 (*i*), 36 (*p*). L'écoulement se fait de droite à gauche: l'ouverture est de 90 millisecondes, l'intervalle non analysé Δt = 6 millisecondes. A la 7ème présentation seulement le texte est entièrement contenu dans la porte.

survient pour qu'on ait l'impression d'un enchaînement, et que le texte semble, à la fois, découpé en fragments et continu. Des temps morts trop longs entre deux présentations rendraient l'analyse inutilisable. La capacité mnésique de notre oreille répond en effet à certaines normes. Le Laboratoire de la Perception d'Eindhoven en avait excellemment tenu compte. Il nous a semblé que la longueur du fragment isolé dans la 'porte électronique' ne devait jamais être inférieure à 90 millisecondes et ne pas excéder 250 millisecondes avec un intervalle de temps de 10 millisecondes entre deux présentations. Un découpage en éléments plus courts que 90 ms — théoriquement il devrait être possible jusqu'à 30 millisecondes — ne permet pas de saisir le contenu, et en outre augmenterait indûment le nombre de présentations. En revanche, si le passage introduit dans la porte est supérieur à 250 millisecondes, étant donné la durée moyenne des phonèmes qui est de l'ordre de 10 à 12 centièmes de secondes, il y aurait donc des moments où le fragment isolé comprendrait, outre l'élément central, une portion trop vaste de l'avant et de l'après nuisant à la perception en quelque sorte 'focalisée' du détail à saisir.

Enfin, comme pour le tracé oscillographique, il y a lieu de faire varier l'ouverture de la porte en fonction du texte et en fonction du locuteur. Du texte, car s'il était invariant pendant 20, 30, 40 centièmes de secondes, il n'y auiait pas intérêt à le parcourir lentement pour entendre des présentations identiques. Si le débit de parole, en revanche, est celui d'un tachylalique, le passage d'un phonème à l'autre est très rapide et la durée du fragment isolé doit en tenir compte. Il nous a semblé que pour la parole, il était bon d'utiliser des portes variant de 100 à 250 millisecondes, les durées de 120 millisecondes à 180 millisecondes étant le plus couramment utilisables, mais il serait regrettable, dans un système de ce genre, de ne pas avoir de possibilité d'investigations en deçà et en delà de ces chiffres. Il en va de même pour l'intervalle de temps entre deux présentations. Il est bon de pouvoir le réduire, pour le parcours de certains objets sonores très instables, à 8 ou même 6 millisecondes — en revanche, il semble qu'un temps plus important que 10 millisecondes soit préjudiciable à la perception de l'ensemble — de même que sur un écran de tube cathodique la rémanence d'une figure doit obéir à certaines constantes de temps liées à la perception visuelle. Les variations individuelles ne sont pas sans limites!

Nous avons pu, à Eindhoven, travailler pendant quelques jours de façon efficace du fait des conditions suivantes, grâce à:

— un compteur électronique de présentations,

— un écran cathodique montrant, inclus dans une porte, le fragment entendu,

— un enregistrement sonagraphique simultané.

TEXTES ENREGISTRÉS

Nous donnons quelques échantillons de textes entendus en tomoanalyse. Ils ont été choisis pour étudier soit des transitions phonétiques, soit des durées, soit des variations mélodiques que nous avions déjà examinées par la technique de l'oscillographie.

Transitions phonétiques

ôte-toi de là [óttwa; dla]
un vase sale [zs]; *une cave fraîche* [vf]
aide-le bien [dlœ], *s'il te plaît* [plè]
de grosses clématites [gr; sklé]
un chapeau doublé bleu [blé, blœ]
un oignon de lis [õdli]
c'est le minimum [sèlmi]
l'huile vierge [lẅ́il vyèrj]
où as-tu mis le pâté de canard [où !as? ouas; tédka]
de la toile lâche [w*àllá*]
il pêche la truite [chla]

Phonétique articulatoire des voyelles nasales françaises.

(1) EN POSITION INITIALE: *ombre*, *onde*; *Indre*, *Inde*; *amble*, *humble*, *angle*, *entre* ...
(2) EN CONTACT AVEC D'AUTRES PHONÈMES dans la chaîne parlée: *donne un bon bain*; *ton compte en banque*; *maman vient*; *viens maman! la pente des monts*; *non content d'arrêter les rayons du soleil*, etc.

Etude des attaques vocaliques en relation avec la durée d'établissement de ces phonèmes:

hâtez-vous; *aidez-vous? arrivez*; *irez-vous? y pensez-vous? habitez-vous là?*
Ces mots ont été prononcés successivement avec une attaque douce

et avec une attaque dure; la durée d'installation du timbre vocalique était régulièrement plus brève dans le second cas.

Étude de la mélodie de la phrase (intonation).

Quel âne! il arrose tout...
C'est une robe ... très longue.
Où as-tu mis le pâté de canard?
Oh! ne tousse pas comme ça!
Oui ... c'est loin, le pont!, etc.

APPLICATIONS A LA PATHOLOGIE DE LA PAROLE

Ce que nous allons en dire est en grande partie simplement déductif.

En effet, si nous avons eu le loisir de pratiquer l'oscillographe dans plusieurs domaines de la pathologie, nous n'avons pas eu la possibilité d'appliquer la technique de la tomoanalyse aux troubles de la parole.

C'est donc seulement sur l'utilisation des oscillogrammes en pathologie que nous pourrons parler à partir de documents.

(1) L'enregistrement direct, c'est-à-dire non stocké sur magnétophone, est souvent impossible, il ne faudra donc pas exiger autant de qualité des documents sonores soumis à l'analyse que s'ils étaient enregistrés 'en direct', mais c'est une nécessité de la clinique: il est évident qu'on peut transporter un magnétophone auprès du patient plus aisément que faire venir le malade auprès des appareils indispensables à la prise d'un oscillogramme, et surtout la vitesse d'enregistrement est plus compatible avec le comportement de gens dont le temps de réaction aux ordres donnés serait trop variable.

Donc, en présence d'états d'inhibition ou de réaction impulsive, il faudra recourir au magnétophone et on choisit sur la bande sonore les portions documentaires les plus intéressantes; ce n'est qu'un pis aller, puisque ce triage qualitatif est fait à l'oreille qui peut laisser échapper des détails.

Ces réserves faites, nous avons pu collectionner des documents oscillographiques concernant le BÉGAIEMENT, les DIVISIONS PALATINES, quelques ANOMALIES DE LA VOIX, des ÉTATS NEUROLOGIQUES (chorée-athétose, parkinson, paralysie générale) et des troubles du rythme, notamment des TACHYLALIES et des BRADYLALIES.

C'est la parole des sourds qui nous a fourni le plus abondamment des documents d'où il ressort les constatations suivantes: VITESSE DU DÉBIT: Les sujets sourds n'ont pratiquement jamais un débit normal: le rythme est troublé du fait de la méconnaissance des accents, de la non perception des rhèses ou ensembles signifiants, des hésitations morphologiques ou syntactiques, du traitement maladroit des groupes phonétiques. La durée globale est généralement allongée de 2/3. Si le sujet arrive à un degré suffisant de fluidité verbale du fait d'une bonne connaissance de la langue, en revanche, il accélère son débit au point d'être parfois incompréhensible.

D'autres facteurs perceptibles à l'oreille sont mis en évidence par ces moyens visuels:

(a) la présence de voyelles surnuméraires qui ne devraient pas être prononcées, ex.: *prèsœkœ*, *jusœka* (*presque*, *jusqu'à*).

(b) des déformations phonétiques par non exécution des modifications usuelles de la chaîne parlée (*une robe peu coûteuse*) [obpe], (*une corde de trop*) [ord dœ] ou par ignorance de la manière de traduire oralement un texte écrit.

(c) des variations extrêmes de la hauteur qui brusquement peut monter, pour certains phonèmes, de plus d'un octave.

L'avantage de ces images est de permettre des mesures chiffrées, et une représentation stable au lieu de la rapide fluctuance que l'oreille entend sans pouvoir en préciser les éléments.

Il serait donc possible, en présence de ces phonèmes fréquemment répétés, de les relever systématiquement et d'établir à leur propos des règles d'intelligibilité.

Les TROUBLES D'ARTICULATION sont également repérables par les modifications des tracés, mais ce sont surtout les grosses perturbations, provoquées par les troubles neurologiques, qui apparaissent avec évidence. On en peut suivre ainsi les transformations, qu'il s'agisse d'aggravation ou de régression du trouble.

C'est donc un apport précis à la clinique qui peut être fourni ainsi par l'oscillogramme.

TOMOANALYSE

Nous n'en avons pas, avons-nous dit, l'expérience dans les cas pathologiques, mais la transposition des erreurs sur un certain nombre de présentations successives, en rendant manifestes les moindres variations de la parole normale et en suggérant en quelque sorte les étapes du mouvement en relation avec les variations sonores apporterait des précisions irréfutables sur les anomalies du mouvement dans l'articulation ou dans l'émission sonore.

Ainsi, dans les divisions palatines, se posent des problèmes non résolus, quant à la rapidité de la fermeture velopharyngée; ni l'endoscope photographique, ni cinématographie ne donnent une précision suffisante pour expliquer certains paradoxes posés par le nasonnement du timbre et par le souffle nasal en des cas où le voile donne, à la radio, un aspect de fermeture. Il est probable qu'il s'agit d'un retard dans l'occlusion ou d'incoordination des mouvements.

TROUBLES DE LA PHONATION ET ANOMALIES DE LA VOIX

Le domaine restreint de l'insuffisance vélaire n'est qu'un aspect dans l'ensemble de la pathologie vocale, et toutes les anomalies du fonctionnement dans le domaine de la phoniatrie, trouveraient dans une exploration systématique par tomoanalyse une description échappant à la subjectivité des appréciations, sans que les autres moyens d'exploration soient, pour autant, négligés. L'appréciation subjective de l'état du sujet, permettrait des contrôles d'efficacité, tant dans le domaine chirurgical, que dans celui de la réadaptation par des moyens médicaux ou rééducatifs.

TROIS PRINCIPES CONCERNANT LE PHONÈME

MILIVOJ PAVLOVIĆ

(Beograd)

LE SIGNE LINGUISTIQUE ET LA STRUCTURE DU PHONÈME DANS LE SYSTÈME DE LA LANGUE

1. La formation du signe linguistique repose sur l'activité articulatoire concernant l'individualisation des phonèmes. C'est le signe linguistique, symbole représentatif, qui se décompose en ses éléments phonatoires. Le mot-phrase, corrélatif du concept psychique, s'approprie la valeur du symbole linguistique, pendant que les éléments phonatoires se stabilisent en qualité de phonèmes, en même temps que les éléments accessoires de force et de tonalité transmettent leurs fonctions spécificatifs aux lexèmes fonctionnels syntactiques. Cet effet analytique de la réalisation de la parole est dû à la translation dans le domaine intellectuel du processus psychophysiologique de la formulation phonique de la pensée. Donc, c'est le contraire à l'état émotionnée qui implique l'expression globale synthétique d'énonciation.

Les phonèmes constituant le mot-symbole ont le caractère complexe: la nature composite du mot-symbole se reproduit dans la nature composite de ses phonèmes. On entend le mot, on le prononce, on le comprend. Dans le même ordre d'idées, c'est aussi le phonème à son tour qu'on entend articulé dans l'ensemble du mot; enfin ce sont encore les fonctions des phonèmes d'un mot qui forment son entité sémantique. — La nature tripartite du mot assure le fonctionnement et le rôle communicatif de la parole dans les contacts sociaux.

Pourtant, les néogrammairiens n'employaient qu'un terme — *Lautbild* (H. Paul, *Prinzipien der Sprachgeschichte* [1909], p. 50ss), représentation du son en général. Ferdinand de Saussure ne distingue que le rapport entre le signifié et le signifiant. "Le signe linguistique", dit-il, "unit non une chose et un nom, mais un concept et une image acoustique. Cette dernière n'est pas le son matériel, chose purement physique, mais l'empreinte psychique de ce son, la représentation qui nous en donne le témoignage de nos sens" (*Cours de linguistique générale*, p. 100). La remarque des éditeurs confirme en réalité que la formule de Ferdinand de Saussure est incomplète. Ce fut la voie qui avait mené l'auteur vers l'idée de séparer la langue de la parole (p. 38ss), ce qui promettait les déviations postérieures, dénuées de réalisme.

Cependant Antoine Meillet, esprit savant positif et clairvoyant, pour éviter les malentendus théoriques et pour assurer l'unité de conception linguistique, croyait utile de stimuler les études des systèmes, des structures et des catégories linguistiques au cours de leur formation. C'est ainsi qu'une série de travaux s'orientaient vers l'étude du langage enfantin. Les observations de O. Bloch, dans son article "Notes sur le langage d'un enfant" (*Mémoires de la Société de Linguistique* XVIII, 37-39) ont été suivies de la thèse de Jules Ronjat *Le développement du langage observé chez un enfant bilingue* (Paris, Champion, 1913), ensuite de la mienne, *Le langage enfantin: Acquisition du serbe et du français par un enfant serbe* (Paris, Champion, 1920). Le caractère linguistique s'affirmait dans ces recherches, dont nous allons citer aussi celles de Karel Ohnesorg (*Fonetická studie o dětské řeči* [Praha, 1948]; *Druhá fonetická studie o dětské řeči* [Brno, 1959] et de Mme Tatiana Slama-Cazacu (*Limbaj și context* [București, 1959]). Cependant, c'est Roman Jakobson qui a traité les imperfections de la parole de l'enfant au point de vue linguistique (*Kindersprache, Aphasie und allgemeine Lautgesetze* [Uppsala, 1941]). A la fin il faut noter aussi le recueil d'articles présentés à la science du langage par les idées d'introduction dues au professeur Marcel Cohen (*Études sur le langage de l'enfant* [Paris, 1962].[1]

[1] Cf. aussi Marcel Cohen, "Sur les langages successifs de l'enfant", *Mélanges*

2. Je reprends ici l'interprétation de la structure du phonème dans l'ensemble du signe linguistique, d'après l'idée générale que j'avais formulée en 1961 (*Zeichen und System der Sprache* [Halle, 1961]; cf. *JF* XXVIII). Cette interprétation sera fondée sur une analyse approfondie, concernant la nature et la structure du phonème aussi bien que sa fonction vu l'ensemble du signe, non seulement dans la synchronie, mais aussi dans la diachronie.

Dans les idées de Saussure, Trubetzkoy (*Grundzüge der Phonologie*) employait le terme 'phonème' pour indiquer la représentation de la valeur phonique de la langue (*Lautbild*). Cependant, ce terme n'a servi que pour constater les corrélations bipartites d'opposition dans le système synchronique de la langue, sans donner aucune explication du processus des changements conditionnés et des déviations diachroniques. La même théorie, basée sur les faits de logique, se tenait en dehors de la dynamogénie psychique et sociale. Pour faire comprendre non seulement les rapports d'ordre structural, mais aussi et surtout les comportements phonatoires des complexes synchroniques et de la perspective en diachronie, il faut caractériser la valeur communicative de la voix et la structure même du phonème.

Les expressions logiques par les cris différenciés, caractéristiques pour les animaux, leur servent aussi de signes communicatifs. On constate les mêmes faits de caractère global chez l'homme: la communication intelligible est fondée sur des éléments phoniques démotivés et différenciés, formant de telle sorte les éléments des signes linguistiques ('mots') lesquels sont en principe les symboles de la réalité: des faits physiques en dehors du 'Moi', et des faits psychiques conçus dans le 'Moi'. Les éléments du signe ne sont pas indépendants, opposés aux valeurs fonctionnelles (prépositions, conjonctions). Inséré dans la phonation, le phonème n'est qu'un élément soumis à l'entité du concept.

Dans sa fonction linguistique, le phonème, non indépendant d'une manière absolue, n'est pas contraire à l'articulation ('son'), qu'il embrasse dans sa valeur, en même temps que la représen-

---

à Vendryès (Paris, 1925); *Idem*, "Sur l'étude du langage enfantin" *Enfance* (Paris, 1952).

tation auditive. Le phonème réunit le complexe conditionné par le principe de communicavité, ses traits composants: passif (audition), actif (prononciation), fonctionnel (intelligibilité). Le phonème, en tant que l'élément du signe, reproduit la structure tripartite, conçue déjà dans la structure du signe linguistique.

3. Le phonème est formé sur la base de la voix reflexe, représentée par des cris ou par "des sons dont le timbre est difficile à déterminer" (Pavlović, *Thèse*, p. 17; cf. R. Meringer, *Aus dem Leben der Sprache* [Berlin, 1908]). Ces éléments phoniques ne le sont pas dans le vrai sens du mot. Cependant on remarque quelquefois certaines émissions de voix avec articulation déjà vers la fin du troisième mois (Meringer), même plus tôt (Ronjat, Pavlović; Stern, *Die Kindersprache* [Leipzig,[4] 1928]). Ces émissions spontanées et fortuites disparaissent à l'époque d'imitation (cf. P. Beck, *Die Nachahmung und ihre Bedeutung für Psychologie und Völkerkunde*).

C'est la première activité psychophysiologique dont l'effet a un caractère passif, résultat de l'audition. Les images sensitives auditives se transforment en images auditives représentatives, provoquant eux-mêmes une activité dynamogène. Les éléments articulatoires fortuits et éphémères de la voix spontanée sont soumis à une espèce de sélection, soutenus par l'audition.

L'image auditive comme un 'tout' global et corrélatif du concept provoque le dynamisme de l'activité motrice également psychophysiologique. Ces réactions ne sont que des tâtonnements: en les corrigeant, on arrive à la stabilisation de l'image motrice constituée d'éléments produits par l'activité des nerfs de l'appareil phonatoire.

Ces tâtonnements se traduisent par des interférences articulatoires avec une tendance vers la corrélation, dans lesquels faits on peut constater les rapports de hiérarchie. Par exemple un enfant serbe distingue *mama*, 'maman' et *nema* 'il n'y a pas', mais il prononce *mama*; quelques semaines plus tard, les deux mots sont corrigés — *mama : mäma*, ensuite *nema*. Les consonnes bilabiales se stabilisent le plus tôt, ensuite les vélaires. Les chuintantes sont au commencement réduites à la frication: *pion* (*pi*^{i}*geon*), *bi*^{i}*u* (*bijou*). La consonne *r* s'oppose le plus longtemps à l'organisation phonatoire. Dans ce sens les observations de K. Ohnesorg sont très instructives, surtout la constatation que l'enfant observé distingue les liquides *r*, *l* (même dans les groupes *kr*, *vr*, *kl*), mais qu'il les reproduit comme une semi-voyelle. C'est l'interprétation de ces faits qui m'a permis de soutenir définitivement la théorie de la nature composite du phonème. L'intégration du phonème correspond à un processus psychophysiologique tripartite.

La nature composite du phonème est conditionnée par le caractère complexe du système cérébral et la distribution fonctionnelle ('localisation') des zones du cortex. La théorie de Troubetzkoy désavouée, nous avons tout le droit de regarder la coordination tripartite dans l'ensemble du phonème en équilibre avec les autres éléments phoniques dans l'ensemble du signe linguistique.

4. Ce sont les études concernant la formation et la structure du langage enfantin qui indiquent et soutiennent la théorie de la structure composite du phonème. Cet équilibre non réalisé chez les sourds-muets et les aveugles, ou bien la structure normale étant désagrégée par quelque accident, on assiste à la réintégration du phonème sur la base des impressions visuelles, respectivement tactiles (enseignement spécial).

La preuve indirecte de la nature composite du phonème, c'est le cas des sourds-muets. Les personnes privées de l'ouïe sont aussi privées de la parole. Elles n'arrivent pas à se former la coordination complète du signe linguistique, non plus la coordination du phonème. Pourtant, on arrive au même but faisant entrer en jeu l'image visuelle, qui soutient aussi la formation du phonème normal. On s'oriente également vers les sensations tactiles (pour faire remarquer les vibrations). C'est de cette façon que les muets arrivent à posséder la parole. Le processus même peut être qualifié comme une espèce de restitution, de réorganisation du phonème.

Un autre type de rectification concerne les traitements spécifiques en vue de corriger certaines imperfections articulatoires, p. ex. bégaiement, perforation de la luette (manque de nasalisation), etc. Les cas de ce genre font remarquer les complexes articulatoires qu'on ne pourrait pas expliquer qu'en les comparant aux processus de l'acquisition du langage enfantin. — Le phonème de la parole des aveugles est organisé normalement, sur la base acoustique, mais sans être soutenu par l'image visuelle, ce qui fait retarder la stabilisation de certains phonèmes, p. ex. l'*u* français.

D'une manière parallèle, le caractère tripartite se reproduit aussi dans la structure du graphème. Cependant chez les aveugles c'est le graphème qui est incomplet. Il doit être restitué sur la base de l'image tactile, dans lequel but Louis Braille a formé les caractères spéciaux par points en relief.

Ces réorganisations, restitutions, rectifications sont dues aux transpositions entre les zones correspondantes du cortex. C'est sur ce principe de transposition qu'on a réussi à représenter les phonèmes par des signes spécifiques visibles composés de traits, de lignes, colorés.

Les auteurs Ralph K. Potter, George A. Kopp et Harriet C. Green nous présentent les résultats de leurs études dans le livre intitulé *Visible Speech* (New York, 1947). L'appareil spectrographe leur a fourni les spectrogrammes qui dévoilent et confirment la complexité différenciée du phonème.

5. Quoique de nature composite, le phonème normal, réorganisé ou restitué, représente un 'tout', l'unité globale d'éléments composants, adaptés aux fonctions passives et actives, opposés au concept, image représentative.

Le principe d'adaptation intervient dans les faits d'assimilation et de dissimilation à contact et à distance. La seconde situation prouve qu'il existe entre les phonèmes d'un 'tout phonatoire' une adhésion concernant la réalisation à venir. Cependant entre les phonèmes il existe une hiérarchie, toujours dominée par la nature des syllabes, ouvertes ou fermées.

L'effet des rapports entre les éléments composants phonémiques internes et interphonémiques est plus important que celui qui concerne le caractère représentatif. Sans avoir compris ce fait, beaucoup de structuralistes n'ont pas manqué de suivre une théorie dénuée de fond. Cependant les signes linguistiques, embrassant les phonèmes composites, appartiennent non seulement à la synchronie, mais aussi à la diachronie. Une évolution dite spontanée offre des déviations causées par l'incohérence de l'audition et de la reproduction (Rousselot, Pavlović, Ohnesorg). Ces déviations sont minimes dans les langues non dérangées par la mixoglottie (p. ex. le lituanien).

Au contraire, ce sont les contacts des langues qui provoquent la désagrégation des systèmes, non seulement les changements phonétiques. La rencontre des phonèmes latins et celtiques *b* : *v*, *v* : *b* (*vini* : *bini*, *berbex* : *verbex*) avec le *v* slave reflète les interférences très intéressantes. Les impulsions articulatoires affaiblies, c'est par le degré *w* que la consonne peut être réorganisée de deux manières: $b > w > v$ ou bien $v > w > b$, même: $b > w > 0$ (cf. mon article, "Orbis pictus"). De même la mixoglottie (lat. $\breve{\imath} > ə > 0$) a provoqué en serbe ou croate le processus ъ/ь > ь^{a} (cf. mon article, *Orbis* VII).

Ce sont également les processus phonémiques qui sont en état de produire les variantes des morphèmes, les déviations morphologiques, p. ex. les formes des pronoms personnels s. dial *gi*, *gim*, *gu*, considérées par A. Belić comme un fait d'analogie (*ga* : *gu* < *ju*). En réalité ce processus n'est pas un fait d'analogie, il est de caractère phonologique. On a constaté *hi* pour *ji* (Code de Novo Brdo, XVIIe s.) aussi en Macédoine (XVe s., B. Koneski). C'est ce fait qui dévoile le transfert réciproque de l'articulation: *ji* < *ih*, *ji* > *hi* et, enfin, la stabilisation de l'articulation plosive *gi*, *gu*.

C'est aussi le cas des consonnes 'parasites', 'prothétiques' qu'il faut expliquer comme un fait phonologique, p. ex. dans le processus mixoglottique, la continuante de la voyelle longue du pronom personnel f. *tâ*, n. *tô* devient brève, mais accompagnée d'un *j*: f. *taj* (*žena*), *toj* (*dete*), ce qui n'est que la décomposition du phonème, donc: $\hat{a} = a + j$, $\hat{o} = o + j$. De même les exemples *lojze*, *grojze* ne représentent pas l'effet de la métathèse, mais celui de la translation de la palatalisation anticipée et phonématisée.

Les processus phonologiques représentent aussi la cause des changements de catégorie du signe linguistique par la décomposition phonématique. L'impératif *gledaj* se transforme en *gle!* (< *gled!*) en s'appropriant la valeur émotive suivie d'une réorganisation phonématique, pour exprimer une exclamation de valeur complexe: il exprime non seulement *pogledaj*, en attirant l'attention de l'interlocuteur, mais aussi l'étonnement, le soupçon, le manque de satisfaction, le rapport opposé.

De la sorte le phonème avec les processus internes des éléments composants, inclus dans l'entité du signe linguistique, se fait valoir dans les faits de phonation synchronique et dans les translations diachroniques. Il prend part au dynamisme linguistique de la synchronie et de la diachronie.

6. "C'est seulement le langage enfantin qui nous donne les indications nécessaires pour comprendre le développement de l'expression orale" (Pavlović, *Thèse*, p. 141). En revanche, les études du langage enfantin sont intégrées dans la problématique linguistique. Dans ce sens, l'analyse psycho-physiologique dévoile trois principes généraux.

(1) D'abord pendant l'acquisition du langage par l'enfant nous assistons aux processus de formation ou plutôt d'organisation de la nature du phonème, corrélative de la nature du signe linguistique. L'entité tripartite (auditive, soutenue par la vue; représentative — concept; reproductive — image motrice), dérangée par les déformations des organes articulatoires, peut être réorganisée ou restituée par des traitements spéciaux.

(2) Le phonème, en tant qu'une entité globale, conditionne les faits d'opposition articulatoire ou tonale, aussi que les faits d'adaptation articulatoire ou tonale. Ce sont les adaptations phonématiques internes qui accompagnent l'organisation du phonème et qui caractérisent le fonctionnement de la langue en général, se reflétant dans la parole des adultes.

(3) Dans la diachronie on ne remarque que les déviations minimes, infinitésimales, remarquables seulement dans la parole des générations à distance (Rousselot). Cependant, dans les cas

de mélange des langues, à la rencontre des systèmes différents, la symbiose linguistique fait remarquer les interférences, les désagrégations de phonèmes, l'extinction de certains éléments phonatoires, les manques de corrélation, même les décompositions phonatoires et l'émancipation phonique, aussi la translation de l'élément palatalisé.

Tous ces faits conditionnées par l'état de virulence prouvent que la parole est englobée dans le mécanisme de l'abstraction de la langue. Les faits et les fonctions de la réalité et de l'abstraction ne font que l'entité linguistique inséparable.

PAPERS OF SECTION MEETINGS

LA RÉALISATION DES GROUPES CONSONANTIQUES CHEZ UN ENFANT TCHÈQUE

LUBOMÍR BARTOŠ

(Brno)

L'étude du langage enfantin en Tchécoslovaquie a une tradition assez riche. Ce fut surtout au cours des deux dernières décades que parurent plusieurs ouvrages monographiques, études et articles dédiés à ces problèmes. Constatons que la plupart d'entre eux ont été conçus et élaborés diachroniquement ce qui se justifie puisqu'ils se sont proposés de suivre les différents phénomènes au cours du développement du langage de l'enfant (tels que le vocalisme, le consonantisme, le vocabulaire, etc.).

Mais il existe un autre procédé également utilisé à cette fin: la saisie synchronique de l'état des réalisations phonétiques à un certain âge, une seule coupe à travers le développement du langage de l'enfant. La présente communication a pour but de démontrer l'état des réalisations des groupes consonantiques chez un garçon tchèque de deux ans révolus.

La réalisation des groupes consonantiques appartient aux stades ultérieurs du développement du langage enfantin. En accord avec la thèse de M. Ohnesorg, on a pu constater que ces groupes n'apparaissent que dans la période où l'enfant, se servant d'un rythme de parole assez rapide, ne prononce plus des syllabes séparées mais des groupes rythmiques entiers. En principe, il se présente trois possibilités: l'enfant peut réaliser les consonnes formant le groupe d'une manière orthophonique, il peut aussi les prononcer d'une manière dyslalique ou, enfin, il peut simplifier les groupes. L'enfant tend à simplifier, dès le début, les groupes de consonnes selon le principe déjà observé par M. Jespersen (au commencement des mots), généralisé par la suite par M. Ohnesorg lequel a enregistré une certaine régularité dans les groupes placés à l'intérieur des groupes rythmiques. D'après ces observations, si la simplification se produit, ce sont les explosives qui, le plus souvent, se maintiennent; si deux fricatives se rencontrent, la plus résistante est celle qui s'est fixée la première. Les premiers groupes consonantiques réalisés correctement sont ceux qui se composent de consonnes considérées comme les plus faciles pour un enfant, telles *m*, *n*, *d*, *b*.

A l'âge de l'enfant auquel s'effectuaient nos observations, la coulée de la parole était si continue qu'on ne pouvait plus noter la décomposition des groupes rythmiques en syllabes. C'est pourquoi on ne distinguait pas si les groupes de consonnes se rencontraient à la suture de la syllabe ou bien à l'intérieur.

En vue d'un classement systématique de notre matériel, on a divisé les consonnes en quatre catégories: (a) explosives, (b) fricatives, (c) affriquées, (d) liquides. (On ne s'est occupé que des combinaisons de deux consonnes; dans les groupes de plusieurs consonnes, très rares d'ailleurs, la simplification se produisait toujours.)

En vertu du classement mentionné ci-dessus, on a obtenu 16 combinaisons représentées, plus ou moins fréquemment, dans les mots constituant le vocabulaire de l'enfant observé. Un grand nombre de combinaisons était formé par explosive + liquide, par fricative + explosive et par explosive + explosive; au contraire, peu fréquentes étaient celles dont l'un des membres était une affriquée. Une fréquence analogue pouvait être aussi observée dans les exemples illustrant ces combinaisons. Ceux-ci étaient très nombreux pour les combinaisons se composant de explosive + liquide, fricative + explosive, etc. Les combinaisons peu fréquentes n'avaient qu'un nombre restreint d'exemples (à l'exception du groupe *čk*).

En résumant le matériel acquis, on peut constater que le nombre total de 102 groupes de deux consonnes est constitué presque par moitié d'explosives (85); ensuite viennent les fricatives (61), puis les liquides (43) et finalement les affriquées (15). En examinant la proportion des consonnes conservées et omises dans ces groupes, on arrive à des constatations intéressantes. Ce sont les affriquées qui se sont maintenues le plus; de quinze affriquées figurant dans les groupes, une seule est tombée. Sur 85 explosives, 72 ont demeuré. La proportion était assez faible pour les fricatives puisque seule la moitié environ (35) est restée. Dans la catégorie des liquides, la situation est la suivante: sur 43 liquides, 11 restent seulement, 17 tombent; le reste oscille.

L'observation des relations entre les consonnes maintenues et omises pendant l'articulation des groupes permet d'arriver à la généralisation des exemples enregistrés. Ayant examiné notre matériel et ayant laissé à part les combinaisons peu fréquentes, on a obtenu 8 règles qui expriment le comportement mutuel des consonnes dans les groupes:

Si dans le groupe se rencontre:

(1) Deux explosives — ou bien elles restent toutes les deux, ou bien la première tombe.
(2) Deux fricatives — la première tombe.
(3) Une explosive et une fricative — ou bien elles restent toutes les deux, ou bien la première, c'est-à-dire l'explosive, tombe.
(4) Une fricative et une explosive (cas inverse) — les deux restent, ou bien la première, c'est-à-dire la fricative, tombe.
(5) Une explosive et une liquide — les deux restent, ou la liquide tombe.
(6) Une liquide et une explosive (cas inverse) — les deux restent, ou la liquide tombe.
(7) Une fricative et une liquide — ou bien les deux restent, ou bien la liquide tombe.
(8) Une liquide et une fricative (cas inverse) — ou bien les deux restent, ou bien la liquide tombe.

Résumons les formules citées. Si dans le groupe se rencontrent deux consonnes de la même catégorie, p. ex. deux explosives, et si l'une tombe, c'est la première qui tombe. Dans les combinaisons très fréquentes d'une explosive avec une fricative et inversement, le même phénomène se produit. Si les deux consonnes ne se maintiennent pas, la première est toujours éliminée; dans le premier cas c'est l'explosive, dans le second, la fricative. Tous les groupes dans lesquels existent les liquides, gardent les deux consonnes, ou perdent l'une d'elles. C'est alors toujours la liquide qui doit céder (sauf le groupe *lh*).

Il résulte de cet aperçu d'ensemble que dans les simplifications des groupes de consonnes on peut constater à juste titre l'existence de certaines lois bien qu'on ne puisse affirmer leur validité pour tous les enfants tchèques.

DISCUSSION:

PAVLOVIĆ: La communication de M. Bartoš est intéressante. Elle fait mieux comprendre le mécanisme articulatoire et les rapports des effets de l'articulation même.

MIKEŠ: Les faits sont intéressants et peuvent être vérifiés dans les matériaux des enfants de diverses langues maternelles. Les faits semblables furent constatés dans le langage des enfants serbocroates.

DAS PROBLEM DER BEZIEHUNG ZWISCHEN DER ARTIKULATIONSBASIS UND DEM PHONOLOGISCHEN SYSTEM IN DER ERFORSCHUNG DER KINDERSPRACHE

JANA DVONČOVÁ

(Bratislava)

Die Erforschung der Kindersprache hat in der Sprachwissenschaft bereits eine lange Tradition. Es ist kein Zufall, dass sich für Fragen der Kindersprache die bedeutendsten Linguisten interessierten und es ist ebenfalls kein Zufall, dass ein ganzes wissenschaftliches Fach — die Pädolinguistik — sich entwickelt hat. Wir sind jedoch der Meinung, dass selbst die Linguistik die Lösung einiger wesentlichen Probleme der Pädolinguistik noch schuldig geblieben ist.

Eine solche Frage, auf die wir da kurzgefasst hinweisen möchten, ist die Frage der Beziehung zwischen dem phonologischen System und der Artikulationsbasis. Der Gegenstand der Erforschung der Kindersprache ist der Prozess des Aufbaus aller Gewohnheiten, die für das Funktionieren des Sprechprozesses und für die Aneignung des phonologischen Systems erforderlich sind, damit es dann zur vollen, allseitigen, sozialen Sprachkommunikation kommen kann. Die Verfolgung dieses Prozesses ist ausserordentlich anspruchsvoll, weil es sich hier um komplizierte gegenseitige Beziehungen und um mehrschichtige Beschaffenheit der Erscheinungen handelt, die an sich wiederum nicht in dasselbe Gebiet der das Kind umschliessende Realität hineingehören. (Ich meine da vor allem die Gebiete der Physiologie, Akustik, Psychologie, die Angehörigkeit zu verschiedenen Sprachsystemen usw.) Die Tatsache, dass diese Problematik nicht durcharbeitet ist, äussert sich als empfindlicher Mangel, dessen Ursache im skeptischen Standpunkt der Sprachwissenschaftler zum Begriff der Artikulationsbasis liegt.[1]

Die Erscheinung selbst, die wir heute als Artikulationsbasis bezeichnen, ist bereits seit drei Jahrhunderten bekannt, zumal wenn wir Wallis' Hinweis "nach Völkern und Landschaften wechselnden kleinen Abweichungen in der Gesamtstellung der Sprachwerkzeuge" der bis heute noch zitiert wird,[2] in Betracht ziehen.

Die Definition des Begriffes Artikulationsbasis, ihre Abgrenzung und Beschreibung haben in den einzelnen Sprachen eine lange Entwicklung durchgemacht. Da sich die Linguisten vor allem, oder auch vielleicht zufällig, für allzu abstrakte Definitionen interessierten, die in gewissem Sinne unvollständig, wobei sie in anderer Hinsicht allzu breit sind, ist es begreiflich, dass sie zu einer gewissen Skepsis gelangt sind. Ich habe vor allem diejenigen Definitionen vor Augen, in denen die Artikulationsbasis als "gemeinsames Bildungsprinzip der Sprachlaute" definiert wird.[3] Unserer Ansicht nach kann man für die Artikulation der Laute der ganzen Sprache kein gemeinsames Artikulationsprinzip finden, weil die Artikulation eine viel kompliziertere physiologische Erscheinung ist, als es dem Sprechenden bei blossem Beobachten scheinen mag. Weiter, wie wir schon darauf in unserem Referat "Die Stellung der Intonation im Rahmen der Artikulationsbasis" am Intonologischen Symposium in Prag im Oktober 1970 hingewiesen haben, betrachten wir als der Wirklichkeit nicht entsprechend, wenn der Begriff der Artikulationsbasis nur auf das

[1] P. Trost, "'Artikulační baze' nebo 'energetická baze'?", *Jazykovědné aktuality; zpravodaj Jazykovědného sdružení při ČSAV* I: 1 (1965). B. Malmberg ist der Meinung, dass der Begriff 'Artikulationsbasis' "un peu impropre" ist (*La phonétique* [Paris, 1962]).

[2] F. Häusler, "Die russische Artikulationsbasis", *Wiss. Z. Univ. Halle — Ges. Sprachwiss.* X: 1 (1961), 249-260. B. Hála,. *Uvedení do fonetiky češtiny na obecně fonetickém základě* (Praha, 1962), 375

[3] H.-H. Wängler, *Grundriss einer Phonetik des Deutschen* (Marburg, 1962), 112: "Darunter versteht man seit Sievers ein gemeinsames Bildungsprinzip der Sprachlaute, das allen Operationen im ganzen (und also auch in Einzelheit) zugrunde liegt".

Gebiet der Lautartikulation begrenzt wird.[4] In der Phonetik können wir eine ziemlich lange Entwicklung der Definition der Artikulationsbasis verfolgen. Diese Entwicklung zeigt uns, wie die Forscher schrittweise über verschiedene Modifikationen der Definition der Artikulationsbasis tiefer und tiefer zum Wesen dieser Erscheinung vordrangen. Auf Grund der Verfolgung der Entwicklung dieser Definitionen sind wir zu folgender eigenen Definition gelangt: [5]

Die Artikulationsbasis ist ein System der motorischen Sprech-Stereotypen der sprechenden Angehörigen einer bestimmten Sprachgemeinschaft.

Das System dieser Stereotypen eignet sich das Kind von seiner Umgebung ebenso an, wie das phonologische System seiner Muttersprache. Dieser Prozess verläuft paralell, er beeinflusst sich gegenseitig und oft bildet er komplizierte Situationen. Wenn wir diese Begegnung verfolgen, stossen wir auf einige vom phonetischen sowie vom linguistischen Standpunkt wichtigen Fragen.

Vom phonetischen Standpunkt sind es diese Fragen:

(1) Ob die Artikulationsbasis eine Gesamtheit von Artikulationsstereotypen darstellt, in der die inneren Beziehungen nicht durch genaue Gesetzmässigkeiten gegeben sind (demgemäss ist es also eine Gesamtheit von zufällig fixierten Bewegungen) oder ob das ein wirkliches durch eigene inneren Gesetzmässigkeiten geregeltes System ist. Diese Gesetzmässigkeiten beeinflussen sich dann gegenseitig mit den Gesetzmässigkeiten des phonologischen Systems und könnten beziehungsweise die Basis für die Gesetzmässigkeiten des phonologischen Systems bilden. Wir sind der Meinung, dass diese motorische Sprech-Stereotypen ein System bilden.

[4] Die Artikulationsbasis einen ausschliesslich physiologischen Begriff darstellt, da der gesamte Sprechprozess als ein vom Nervensystem gelenkte Synergismus der Muskeltätigkeit realisiert wird. Es ist also erforderlich, in diesen Begriff auch die Intonation und andere suprasegmentale Erscheinungen einzuschliessen, da sie auch ihre eigene physiologische Spezifik haben. Stimmabläufe können in dieser Hinsicht von Artikulationsabläufen nicht getrennt werden.

[5] J. Dvončová, "Einige Bemerkungen zum Begriff Artikulationsbasis", X. Sprechwissenschaftliche Fachtagung, Erfurt 1968 (Referat); J. Dvončová, "Artikulačná báza", *Jazykovedné štúdie* XI (Bratislava; 1971), 288-295.

(2) Man kann auch noch die Frage zulassen, ob der Terminus 'Artikulationsbasis' für diesen Begriff am besten geeignet ist. Wenn dieser Terminus, in dem Sinne wie er oben definiert wurde, abgegrenzt wird, erfüllt er, unserer Ansicht nach, seine Funktion ziemlich gut und er sollte, unter der Berücksichtigung einer gewissen traditionsmässigen Stabilisiertheit, beibehalten werden, wenn auch seine gegenwärtigen Definitionen ziemlich unterschiedlich sind.

Vom weiteren linguistischen Standpunkt handelt es sich um folgende Fragen:

(1) Wie ist die Beschaffenheit der Gesetzmässigkeiten der Artikulationsbasis in Bezug auf die Gesetzmässigkeiten des phonologischen Systems? Sind sie grundsätzlich vergleichbar? Wie ist ihre gegenseitige Wirkung? L. Zinder [6] hebt hervor, dass die Artikulationsbasis die Richtung der Lautevolution bestimmt und die Festigkeit (Kompaktheit) des Lautsystems aufhellt. Obzwar aus dieser Behauptung Zinders nicht die Schlussfolgerung hervorgeht, dass es sich da um eine vollständige Abhängigkeit des phonologischen Systems von der Artikulationsbasis handelt, kommt jedoch die Wichtigkeit des Einflusses der Artikulationsbasis auf das phonologische System klar zum Ausdruck. Wenn wir in Betracht ziehen, dass jedes Glied dieses Paares sowohl mit dem Sprechenden als einem Individuum als auch mit der kollektiven Norm der Angehörigen derselben Sprachgemeinschaft verbunden ist, sehen wir durch welche komplizierte Beziehungen die beiden Glieder verbunden sind. Wir fragen: Welches Glied ist dann vom Standpunkt der Sprache und welches vom Standpunkt des sprechenden Individuums dominant? Welches Glied ist dominant vom Standpunkt der Entwicklung und welches vom synchronischen Standpunkt? Handelt es sich da um eine wechselhafte oder um eine unveränderliche Dominanz?

(2) Eine weitere wichtige Frage ist das Klarlegen der Beziehung zwischen der Artikulationsbasis und der ortoepischen Norm. Das Bewusstsein der Bedeutsamkeit dieser Frage ist nicht neu.

[6] L. Zinder, *Obščaja fonetika* (Leningrad, 1960), 79.

Auf diese Frage macht z.B. H. Krech [7] aufmerksam, der zur Ansicht kommt, dass "auch als Ausgangspunkt für eine einheitliche deutsche Lautung die Feststellung und Abgrenzung der erforderlichen Artikulationsbasis ein Grundproblem bildet". Wir halten es für möglich, folgende Schlussfolgerung zu ziehen: Nichts, was nicht in Übereinstimmung mit der Artikulationsbasis wäre, darf Bestandteil der ortoepischen Kodifikation werden, weil es keine Aussicht hat, ein bestehender Wert der ortoepischen Norm zu bleiben.

Die praktische Tragweite einer gründlichen theoretischen Lösung dieser Fragen für den Unterricht der Kindersprache, für die Logopädie, Phoniatrie u.s.w., für den Fremdsprachenunterricht sowie für die Lösung einiger orthoepischen Probleme, braucht man nicht besonders hervorzuheben.

Zur Illustration der angedeuteten theoretischen Problematik wollen wir nur einige Beispiele aus dem slowakischen Sprachmaterial anführen. Eine der gewöhnlichsten Vorstellungen ist, dass die Stabilisiertheit des phonologischen Systems im gleichen Masse der Stabilisiertheit der Artikulationsbasis entspricht. Auch wenn diese Ansicht nicht ausgesprochen auf diese Weise formuliert wird, kann man sie in den Beschreibungen der Artikulationsbasissen einzelner Sprachen verfolgen, die in mancher Hinsicht viel mehr Beschreibungen der phonologischen Systeme als deren Artikulationsbasissen sind. Über die Stabilisiertheit des Inventars des slowakischen phonologischen Systems der Gegenwart gibt es keinen Zweifel. Wenn wir genau die Aussprache einzelner bestimmten Phonemen entsprechender Laute verfolgen, stellen wir fest, dass es in mehreren Fällen zwei deutliche Aussprachevarianten gibt, die, unserer Erfahrung nach, eine überindividuelle Gültigkeit haben. Diese Varianten weisen auf die Unstabilisiertheit der slowakischen Artikulationsbasis hin. Die Ursache kann einerseits die Nichtbeendigung der Konvergenz der mundartlichen Aussprache sein, doch andererseits auch die Begegnung der Artikulationsbasissen grösserer Sprachgruppen, die durch ihre Wirkung

[7] H. Krech, "Zur Artikulationsbasis der deutschen Hochlautung", *Zeitschrift für Phonetik u. allgemeine Sprachwissenschaft* 7 (1954), 95.

auf das Gebiet der Slowakei über Gebiete der Artikulationsbasissen hinausreichen (Abb. 1a, 1b).

Zu dieser Situation trägt auch die zentrale geographische Lage zwischen den übrigen slawischen Sprachen bei. Nicht zu letzt ist es auch die nationale Buntheit in einigen Gebieten der Slowakei. Im Slowakischen haben wir z.B. neben der Aussprache des Vokals *a* als niederen, mittleren Vokal (Abb. 1a,b) [8] auch eine mehr

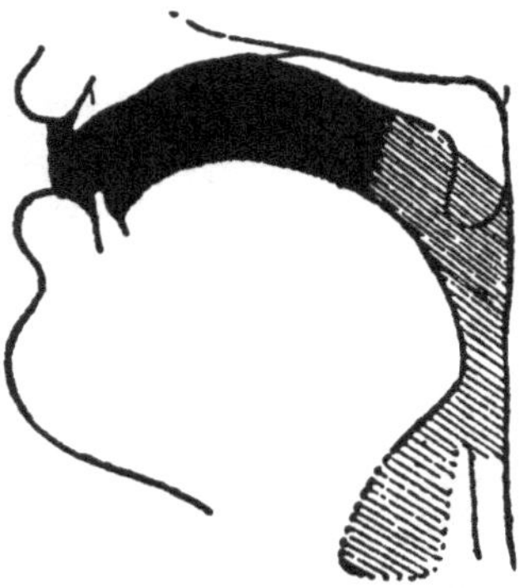

Abb. 1a (nach J. Stanislav).

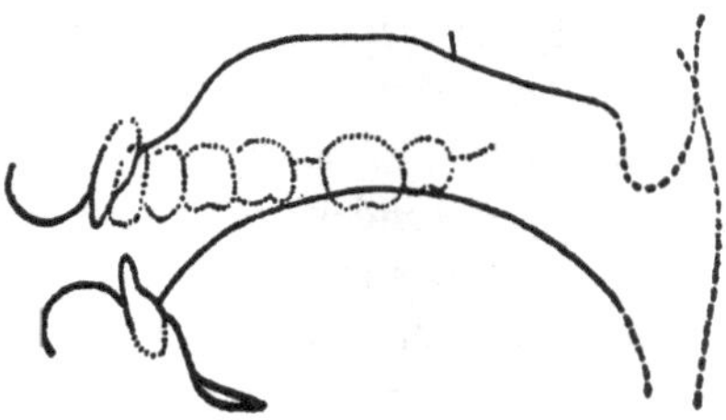

Abb. 1b (nach B. Hála).

velare Aussprache festgestellt (Abb. 2). [9] Diese Aussprache ist nicht mit dem aus dem Ungarischen ins Slowakische unrechtmässig übertragenen velaren *a* identisch (Abb. 3).[10] Ähnlich gibt es auch bei der Artikulation palataler Konsonanten *ť, ď, ň, ľ* zwei nebeneinander bestehende Artikulationsvarianten. Die eine ist dem palatalisierten *t, d, n, l* ähnlich und wird mit der Zungenspitze am Gaumengewölbe artikuliert. Auch das Feststellen der Artikulationsstelle ist in diesen Fällen problematisch, weil das keine eindeutig palatale Laute sind. Im Atlas der slowakischen Laute [11]

[8] J. Stanislav, *Slovenská výslovnosť* (Martin, 1951), 115-116. B. Hála, *Základy spisovné výslovnosti slovenské a srovnání s výslovností českou* (Praha, 1929), 80.

[9] J. Stanislav, 1951: 116; J. Dvončová, G. Jenča, und Á. Kráľ, *Atlas slovenských hlások* (Bratislava, 1969), Tab. 2.

[10] J. Molnár, *A magyar beszédhangok atlasza* (Budapest, 1970), 52.

[11] J. Dvončová, G. Jenča, Á. Kráľ, *Atlas slovenských hlások* (Bratislava, 1969).

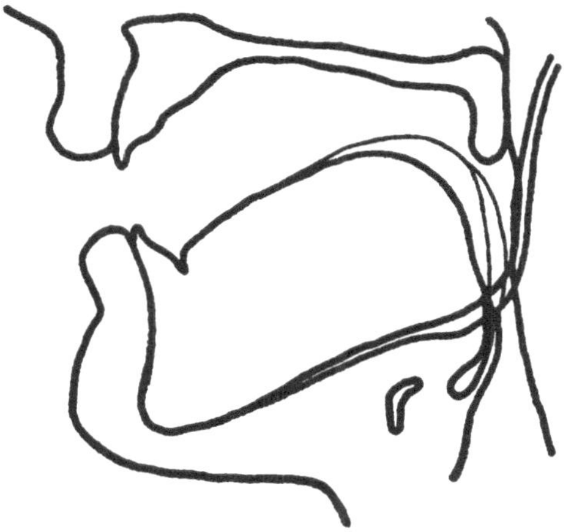

Abb. 2 (nach Dvončová-Jenča-Král).

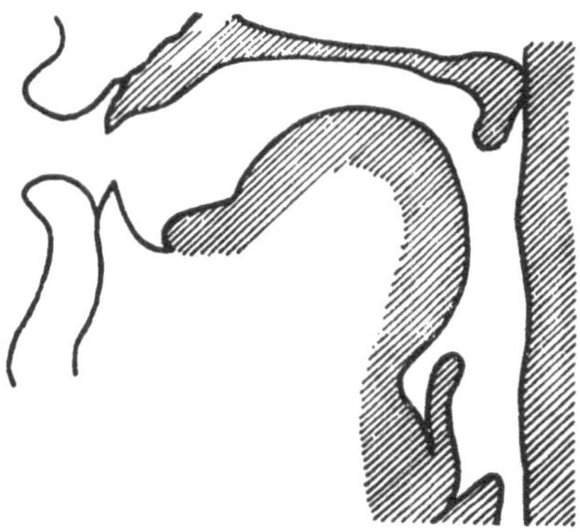

Abb. 3 (nach J. Molnár).

haben wir den Terminus alveopalatale Konsonanten verwendet. Die zweite Variante ist die Artikulation mit der Zungenspitze hinter den unteren Zähnen bei gleichzeitiger Berührung des Gaumens mit dem vorderen Teil des Zungenrückens. Diese Aussprache ist ausgeprägt palatal. Ein anderes Beispiel: bei der Artikulation der sonoren Laute *r*, *l* und des alveolaren Nasals *n* tritt neben der Artikulation mit dem Rücken der Zungenspitze sehr oft die Artikulation mit einer ausgeprägt kakuminaler Stellung der Zungenspitze. Diese führt in einigen Fällen zum Ergebnis, dass sich die Artikulationsstelle deutlich von den vorderen Alveolen zur Scheide zwischen den vorderen und hinteren Alveolen oder sogar bis zu den hinteren Alveolen verschiebt. Auch in diesem Falle ist die Feststellung der Artikulationsstelle problematisch. Diese und auch andere Fälle aus dem Gebiet der Zischlaute habe ich bereits während der vorbereitenden Arbeiten zum Atlas der slowakischen Laute in einer Sonderstudie beschrieben.[12]

Wir sind der Meinung, dass die herrschende Unklarheit in der Beziehung zwischen der Artikulationsbasis und dem phonologischen System bei der Erforschung der Kindersprache zur Folge hat, dass die Forschung überwiegend auf der phonologischen Ebene erfolgt. Das bedeutet jedoch eine gewisse Verarmung der ganzen Problematik, weil gerade beim Kind die Möglichkeit gegeben ist, das Stabilisieren der Artikulationsbasis zu verfolgen und dadurch leichter auch zum Wesen einiger ihrer Gesetzmässigkeiten vorzudringen. Beim erwachsenen Sprachbenutzer finden wir eine stabilisierte wenn auch bisweilen modifizierte Artikulationsbasis. Wenn ein Logopäde in seiner Praxis auf gewisse Abweichungen in der Kinderaussprache stösst, kann er auf Grund mangelhafter Kenntnisse gerade diejenigen Elemente der Aussprache eines Lautes ändern, die produktiv und für die slowakische Artikulationsbasis typisch sind.

Wir schliessen also wieder mit der Behauptung ab, dass der skeptische Standpunkt der Sprachwissenschaftler zum Begriff Artikulationsbasis und das Umgehen der Lösung der Beziehung zwischen der Artikulationsbasis und dem phonologischen System die Erforschung der Kindersprache erschwert und manchmal auch zu unnötigen Missverständnissen in der Interprätation des experimentellen phonetischen Materials führt.

[12] J. Dvončová, "Z experimentálniho výzkumu slovenské výslovnosti", *Jazykovedný časopis* 15 (1964), 138.

СЛОВООБРАЗОВАТЕЛЬНЫЕ НЕОЛОГИЗМЫ В ДЕТСКОЙ РЕЧИ НА МАТЕРИАЛАХ ПОЛЬСКОГО И РУССКОГО ЯЗЫКА

МАРИЯ ХМУРА-КЛЕКОТОВА

(Warszawa)*

Целью нашей работы является анализ:

(1) условий, в которых возникают аналогичные словообразовательные формации в детской речи,
(2) механизмов возникновения детских неологизмов,
(3) очерёдности появления отдельных словообразовательных категорий в этих неологизмах.

Наши рассуждения основаны на материале, охватывающем около 5 000 польских и около 1 000 русских примеров лексикальных неологизмов в детской речи, то есть слов, не заимствованных из речи окружающих, а созданных ребёнком самостоятельно по образцу словообразовательных формаций, существующих в языке.

Польский материал был собран при помощи: (1) метода постоянного наблюдения, охватывающего период нескольких (в некоторых случаях даже более чем десяти) лет, польских детей. Эти наблюдения велись в течение 1887-1904 (дневники речи пятерых детей Я. Бодуэна де Куртенэ), а также в пятидесятых и шестидесятых годах нашего столетия (6 дневников речи, разработанных под руководством проф. С. Шумана, 3 дневника, которые велись по инициативе автора лицами со специальностью польская филология и лингвистика,

* Not present in person.

а также один, который вела сама автор в течение 6 лет. В четырёх последних дневниках наблюдатели проявляли специальный интерес к неологизмам), (2) разработанной автором словообразовательной анкеты, при помощи которой было исследовано 80 детей в возрасте от 3 до 7 лет.

Примеры из русской детской речи взяты из (1) труда А.Н. Гвоздева (см. библиографию), а также (2) из книги К. Чуковского (см. там же), Исходными для анализа являются польские материалы, в первую очередь данные, собранные автором при наблюдении за речью сына (1147 примеров), так как материалы по русской детской речи были собраны с несколько иными целями, и поэтому они не содержат всех необходимых данных, например К. Чуковский указывает возраст детей с точностью до полугода (а часто возраст вообще не указан), тогда как в труде Гвоздева почти отсутствует информация о том, при каких обстоятельствах появились анализируемые аналогичные формации.

Словообразовательная классификация анализируемых примеров проведена в соответствии с принципами, разработанными профессором В. Дорошевским, с одновременным учётом основных отличий, применяемых М. Докулилом (см. библиографию).

УСЛОВИЯ ВОЗНИКНОВЕНИЯ НЕОЛОГИЗМОВ

Анализ ситуации, в которой в высказываниях ребёнка появляется неологизм, позволяет утверждать, что непосредственный импульс к созданию нового слова может исходить: (1) от окружающей ребёнка внеязыковой действительности — наблюдение за новыми предметами, признаками, функциями, которые не имеют ещё соответствующего названия в запасе слов ребёнка, (2) от потребности выражения своего отношения к этой действительности, а также (3) от предыдущего высказывания взрослого или самого ребёнка (восприятие нового, ранее неизвестного слова, размышление о составе знакомых и вновь познаваемых слов).

Второй фактор выражается в создании экспрессивной формации, например: Г. (3; 8, 4): *już nie ma tej wstrętnej Gęsiówy!* (Gęsiowka, развалины прежней тюрьмы и тюрьма, разобрана в 1963 году; там была котельная, которая сильно дымила). *Смотри, какую красоту я делаю, какую красотищу, какую красотину! Смотри, какая красота! Ж.* (3; 10, 25) Гв. 300* и не всегда играет роль в создании других неологизмов, тогда как первый и третий факторы являются необходимыми условиями создания нового слова, причём в конкретных случаях один из них выдвигается на передний план: например Г. (2; 3, 31) (вернувшись с кладбища): *pan dzwoniarz był na cmentarzu.* У меня курица окурилась. Ж. (3; 9, 25) Гв. 301. М. (обращаясь к Г.): *ty paskudniku*! Г. (3; 5, 7): *a ty paskudniczka! Почему кровати называют, а не спалилы? Тогда бы все понимали, что на них спят. Ж.* (7; 8, 31) Гв. 332.

Неологизмы возникают чаще всего в условиях интеллектуального и эмоционального возбуждения, сопутствующего перемене окружения, путешествию, интересной тематической игре, рассматриванию новых картинок и иллюстраций и т. д. Ребёнок испытывает в такой ситуации потребность вслух назвать свои наблюдения: неизвестные предметы, их признаки, новые специфические действия. Такое возбуждение может быть вызвано также разговором, чтением книжки, передачей по радио или телевидению, во время которых ребёнок сталкивается с новыми, неизвестными словами и анализирует их путём соотношения к известным ему словам и соответствующим представлениям.

ПСИХИЧЕСКИЕ ИСТОЧНИКИ ВОЗНИКНОВЕНИЯ НЕОЛОГИЗМОВ

Источников неологизмов следует искать в определённых чертах детской психики, свойственных детям на данном этапе развития. Назовём некоторые из них:

* В дальнейшем в тексте приняты следующие обозначения: Гв. обозначает труды Гвоздева, Ч. — Чуковского, цифра, следующая за буквой, обозначает страницу. Цитируя примеры, мы приводим инициалы ребёнка и его возраст в следующей очерёдности: годы, месяцы, дни.

(1) Сильная потребность называть вслух свои наблюдения при общении со взрослыми (возраст вопросов) и во время монологов в процессе игры; “называние” облегчает ребёнку анализ и понимание окружающей действительности, делает её более доступной, что имеет большое значение для его общего развития.
(2) Незнание многих названий и понятий (небольшой запас слов), а также недостаточное овладение запасом, диапазоном употреблений и функций словообразовательных формантов при одновременном растущем осознании самого факта делимости слов и происхождения одних слов от других; выражается это с одной стороны в неологизмах, в которых ребёнок называет предмет, исходя из иного, чем в речи взрослых, признака, а именно из признака, доминирующего в восприятии ребёнка, например *trzymadło* ‘ручка от сковороды’ Г. (2; 3, 19), *podpieraczka* ‘перила’ Г. (2; 10, 2), или для которых в речи взрослых существует заимствованное название, например *zaświecacz* ‘электрический контакт’ Г. (2; 6, 22); с другой стороны в неологизмах, отличающихся от общеупотребительных названий формантом: например *tramwajny* ‘остановка’ Г. (2; 7, 25), вместо *tramwajowy*, *нефтной вагон* Ж. (4; 4, 17) Гв. 309, *почтаник* (2, 5) Ч. 15, *копатка* (‘лопатка’) (3, 0) Ч. 15.
(3) Неотдавание себе отчёта в существовании в языке постоянных названий. Ребёнок начинает осознавать етот факт лишь впоследствии, что постепенно задерживает процесс создавания новых слов. Отсутствие этого сознания выражается в поспешности и лёгкости, с которой дети дошкольного воазраста пользуются разнообразными средствами образования производных слов.
(4) Большая эмоциональная возбудимость, которая обуславливает создание многочисленных экспрессивных формаций.

Четыре вышеприведённые фактора являются причиной того, что детская речь отличается склонностью к созданию импровизированных названий исполнителей, инструментов и предметов действия, экспрессивных субстантивированных прилагательных (типа атрибутивное существительное—n. attributivum), увеличительных, уничижительных и уменьшительных прилагательных, образованных от существительных, а также глаголов и прилагательных, образованных от существительных (выражающих разнообразные взаимные соотношения между предметами), как и многочисленных префиксальных глаголов.

(5) Своеобразное понимание соотношения между названием и его предметом — частые вопросы о том, почему данный предмет так называется, являются вероятно следствием убеждения, что существует тесная связь между признакамн предмета и его названием. В результате такого отношеиия к названию, как к одному из признаков предмета, ребёнок критикует неоднократно название, с которым о нпознакомился, и создаёт новое, по его мнению лучшее: М.: *poczęstuj się herbatnikiem.* Мись (4, 3): *to nie herbatniki, to kawniki, bo przecież jemy je z kawą, nie z herbatą*; заинтересованность ребёнка вопросами языка, относящаяся к этой группе явлений, выражается кроме того в поисках этимологии для уже известных и новых слов, делении на составные значащие элементы и размышлении о их значении; примеры: М. (Михалу): *nie bądź taki przemądrzały.* М. (3; 6, 0): *dlaczego właściwieprze- ?* Г. (3; 0, 1): *dlaczego nazywa się Lwów, czy tam były lwy ? Почему перчатки, надо пальчатки.* Ч. 46. *Деревния — где деревьев много?* Ж. (4; 7, 3) Гв. 309. *Не художник, хорошник!* Ч. 43.

(6) Поиски этимологии слов связаны с характерной для этого периода развития потребностью обоснования всего, с чем ребёнок сталкивается. Как свидетельствуют наблюдения психологов, в этот период ребёнок задаёт взрослому много вопросов, начинающихся с “почему” (знаменитое “сто тысяч почему”), причём в ответ на эти вопросы ожидает определения причины или цели (предназначения) данного предмета или явления и т.п. формирующееся при помощи наблюдений и собственного опыта причинно-следственное мышление часто носит в этот период причинно-целевой характер; причиной

определённого положения вещей бывает конкретная цель, на достижение которой направлена человеческая деятельность. Понятия причины и цели, не отделённые чётко, проявляются в процессе создания новой словообразовательной формации: предмет наблюдения получает название как с точки зрения его предназначения *wycieraczka* — do *wycierania*, *herbatniki* — для еды с *herbatą* ('чаем') *solniczka* — для хранения *soli*), так и исходя из определённых его признаков *szarak* jest *szary*, *nauczyciel naucza*). С точки зрения языка "обосновывает", мотивирует с одной стороны словообразовательная основа (т. наз. обосновывание), с другой стороны — слово или группа слов, относящиеся к тому же словообразовательному типу, в которых тот же формант выполняет ту же функцию (т. наз. мотивирование). В своём этимологизировании дети ищут в первую очередь мотивирования первого типа и обычно ограничиваются нахождением его для новых слов. В результате этого процесса создаются часто такие неологизмы, в которых созданная по принципу аналогии правильная словообразовательная основа является ошибочной, поскольку функция форманта была понята неправильно — речь идёт о заимствованных словах и т. п.; примеры: Г. (3; 1, 20): *co to za pani*? М.: *pani portierka* Г.: *a co ona robi*? *portieruje i tak pracuje*? Г. (4; 6, 18): (интерпретирует новые для него имена и фамилии): *Szymek to Szym*, (4; 4, 15): *pan Mrożek to Mróg*.

(7) Одна из характерных черт восприятия, издавна отмеченная психологами: при сравнении наблюдений легче замечаются различия, чем сходства, поэтому ребёнок раньше замечает контрастные, противоположные признаки сравниваемых объектов и их лексикальные показатели. Противопоставляются значащие морфемы слов, в первую очередь словообразовательные основы и такие префиксальные и суффиксальные форманты, которые имеют чёткое, конкретное значение (показатели грамматического рода, уменьшительные и увеличительные форманты существительних, префиксы, выражающие противоположные направления и их способы действия, в глаголах и отглагольных существитель-

ных). Таковы же результаты наблюдений, сделанных К. Чуковским, который пишет в частности:

> “Среди многочисленных методов, при помощи которых ребёнком усваивается общенародная речь, смысловая систематизация слов занимает не последнее место. По представлению ребёнка, многие слова живут парами; у каждого из этих слов есть двойник, чаще всего являющийся его антитезой... Эти словесные пары, насколько я мог заметить, являются для ребёнка не только двойниками по смыслу, но в большинстве случаев — по звуку”. (Ч. 163)

(8) Дух противоречия (эмоционального и интеллектуального), свойственный этому периоду развития, который находит выход в противопоставлении слов, о которых речь шла выше.
(9) Эгоцентризм восприятия и мышления ребёнка: к окружающему миру ребёнок подходит с точки зрения собственных потребностей, собственной деятельности, придаёт ему человеческие черты, лучше вего известные из наблюдения за окружающими. Это выражается в оживлении мира предметов и вещей (например, путём наделения их чертами человеческого рода), а также в огромной “популярности” определённых словообразовательных категорий, например отглагольные объекты действия и действующие лица (nomina instrumenti и nomina agentis). Примеры анимизации: Г. (4; 9, 21): *pan dyń i pani dynia*; Сл. Б. (3, 2): *panna jagoda i pan jagód* Г.Д. (4, 0) матери: *daleko od Falenicy jest Świdra*? М.: *Świder nie Świdra* Г.: *a ja mówie Świdra, bo to jest dziewczynka. Woda zawsze jest dziewczynka. Москва — она, Пенза — она, Ростов — он.* Ж. (7; 6, 27). ГВ. 330.
(10) Потребность в игре: ребёнок играет звучанием слов, подбирая рифмы и меняя с этой целью словообразовательную структуру слов, точнее их конечных слогов.

МЕХАНИЗМ ВОЗНИКНОВЕНИЯ НОВОГО СЛОВА В ДЕТСКОЙ РЕЧИ

Исходя из способов возникновения неологизма и факторов, определяющих его форму, материал, полученный в результате наблюдений, классифицируется следующим образом:

(1) В основе определённой группы новосозданных слов (почему иногда трудно назвать их неологизмами, будет ясно из приведённых примеров) лежит неправильное деление на морфемы, ошибочное выделение слов в предложении, а также неправильное понимание функции выделенного словообразовательного форманта.

Примеры: М. (Гжесю): *uważaj, bo para bucha.* Г. (2; 2, 24), дуя: *Grześ wypędzi parabuchę.* М.: *mówiłam od rzeczy.* Г. (4; 5, 16): *a co to jest odrzecz?* М.: *znowu będziesz kopcił, nie chcę i tak tu duszno.* Г. (4; 0, 25): *a ja wolę w dusznie siedzieć.* К этой группе относятся также примеры: *Szym, portieruje.* В формации *Szymek* функция форманта *-ек,* понятая ребёнком слишком узко, ведёт к установлению в качестве словообразовательной основы чистого корня по аналогии таким парам слов, как *domek -dom, kotek -kot*, тогда как в рассматриваемом примере мы имеем дело с т. наз. дезинтегральным производным словом, созданным путём "обменного" образования производных слов от словообразовательной основа *Szymon.* В формации *portierka* ребёнок усматривает структуру действующего лица (nomen agentis) и ищет отглагольную словообразовательную основу, тогда как слово это является вероятнее всего женским зквивалентом существительного *portier*, которое, как известно, заимствовано из французского языка (и связано с существительным porte — 'дверь'). *Муж стрекозы — стрекозёл*, Ч. 11; *трещётки — три тётки*, Ч. 26; *двояшки — два Яшки*, Ч. 59.

(2) Другая группа неологизмов может быть объяснена, как контаминация двух близких по значению слов, с недостаточным выделением звучания префиксальных и суффиксальных морфем.

Примеры: Г. (2; 2, 9): *a ta woda jest gotowana?* М.: *tak, woda sodowa jest zawsze gotowana.* Г. ('радостно'): *woda sodowana, daj mi wody sodowanej.* М. ('гасит свет'), Г. (2; 6, 1): *dla ... dlaczemu zgasiłaś światło? Мапин* ('мамин + папин'), Ч. 38; *волосетка* ('сетка для волос'); *отмухиваться* ('отмахиваться от мух'), Ч. 37.

Недостаточная дифференциация суффиксов является причиной таких неточных воспроизведений слышанного, как *czapka rogatka* (2; 4, 4) (несколько часов тому назад, осматривая памятник Килинского, ребёнок слышал название *rogatywka*), или *lokawka* (2; 6, 11), (вместо употребляемого мною названия *lokówka*, 'бигуди').

Примеры смешивания суффиксов (в глаголах и отглагольных существительных): Г. (2; 5, 11) (о телевизионном спектакле): *panie tańczą i wybierają nogami*; (2; 10, 7) (во время мытья, опасаясь неосторожного движения матери): *ale nie zarażaj, nie porażaj mi bródki* (вместо *nie uraź*); Г. (4; 5, 10) (вспоминая слышанное ранее стихотворение): *a kot z odziwu prawie oniemiał* + М.: *jak*? Г. (отдавая себе отчёт, в каком месте воспоминание неточно, исправляет): *z zadziwu, zadziwił się*. Из вышеприведённых примеров видно, как ребёнок, воспроизводя ранее слышанные слова, безошибочно воспроизводит их главную значащую часть, но смешивает префиксальные и суффиксальные морфемы, к которым он отнёсся видимо при запоминании, как к менее важным, в результате чего они слабее запечатлелись в его памяти. Ища сознательно или подсознательно этимологию для неясных со словообразовательной точки зрения слов, ребёнок иногда, (как это часто происходит в говорах), преобразует их фонетически. Таким образом *fasolka* преображается в *kwasolkę*, Мись, (2; 3, 21), из *faszerowanej* капусты выходит *maszerowana*, В. (5; 1); Г. (3; 10, 28): *dlaczego panie obożne, bo obozują*? (имеет в виду *woźne*). Много подобных примеров из речи русских детей цитирует в своей книге К. Чуковский: *мазелин* — вазелин, *теплометр* — термометр, *колоток* — молоток, Ч. 24-29.

(3) Значительна группа неологизмов, в которых ребёнок акцентирует значение составных значащих частей слова, противопоставляя их себе в двух словах. Одно из этой пары слов чаще всего известно и часто употребляется, тогда как другое является неологизмом, хотя случается и так, что оба слова являются неологизмами. В основе такой реакции на слова лежит известная характерная особенность раннего

детского восприятия, о которой мы уже упоминали, а именно лёгкость улавливаниях различий и контрастов. Противопоставляются с одной стороны содержание, выраженное при помощи основной значащей части слова, а с другой стороны префиксальные и суффиксальные форманты. Часто неологизму, основанному на таком противопоставлении формантов, сопутствует слово, досказывающее, разъясняющее значение форманта, разницу в значении между двумя сравниваемыми формациями. Соотношение противопоставления выражается при помощи союзов *a*, *ale*, *nie.....tylko*, хотя бывает, что оно не находит словесного выражения. Противопоставляемые форманты отличаются легко уловимой значащей функцией, модифицирующей значение словообразовательной основы (уменьшительной -deminutiva, увеличительной -augmentativa и ухудшающей -deteriorativa, отличающей названия существ женского рода от существ мужского рода и мужского от женского в существительных, образованных от существительных, модальные значения, временные и пространственные определения, выраженные при помощи префиксов и суффиксов глаголов). Многочисленны такие примеры отрыва частицы *nie* (и близкого по значению предлога *bez*) в словах, содержаших отрицание.

Примеры: М.: *rozlejesz wodę i taki będzie skutek*! Г. (2; 2, 25): *skutek mały*, *duży skut*. Г. (3; 1, 11): *aż do sufitu drzaza*! ‘*drzazga*’. Г. (2; 11, 18): *to jest łóda, bo duża*. Г. (2; 2, 28): *to nie sarna, to jest sarn*. Г. (2; 3, 7) (перед телевизором): *teraz pani syrenka*, (при виде диктора): *a to pan syrenowy*. Б. (Гжесю): *Zosia-samosia*. Г. (4; 8, 28): *tylko nie samosia*, *samoś najwyżej*, *albo samek*. М.: *ty jesteś szpaś*. Г. (3; 0, 1): *a ty szpasia*, *szpasica*. М.: *mój kochany rekrut-Bum*. Г. (3; 7, 19): *moja kochana rekrut-Bumka*. М.: *aleś ty fujara*! Г. (3; 4, 0): *a ty fujarka*! *Притонула и вытонула,* Ч. 20; *я черепах, не черепаха,* Ч. 36; *мастер стал сработный, он теперь не безработный, у него есть работа,* Ж. (7; 8, 8) Гв. 333; *чаянно — нечаянно,* Ч. 34; *ненавижу — навижу,* Ч. 34. Необыкновенно богато представлены противопоставления значений префиксов в глаголах и отглагольных существительных. Значение префикса подчёркивается также путём его отрыва от корня (инверсия), или при помощи добавления наречий, досказывающих и разъясняющих его значение. Примеры: М.: *do ciepłej wody w wannie treba się przyz wyczajać*; Г. (3; 4, 13) (раздетый и поставленный в ванну): *a teraz będziesz mnie zwyczaić? zwyczaisz mnie, prawda?* Г. (3; 0, 12): *dodmucham materac, a teraz wydmucham* — 'выпушу из него воздух'. Г. (2; 8, 15): *nacisnąłem pedał, a teraz odcisnąłem* 'перестал нажимать'. Г. (3; 1, 3): *one, te kamienie się wypłaszczyły, a nie nadpłaszczyły tylko.* Г. (2; 9, 26) (матери): *przytwierdź to ... a teraz muszę odtwierdzić* — оторвать. Г. (2; 2, 10): *podgasić trzeba gaz, nie zgasić.* Г. (2; 9, 1): *to nie spychacz, tylko podpychacz.* Г. (3; 9, 26): *przytuliłem się do ciebie, a teraz odtuliłem.* Г. (3; 0, 28): *teraz trochę przyzimniało?* Г. (4; 0, 2): *zszyj mi, bo mi się kołderka rozszyła.* Г. (2; 9, 14): *tu jest zjeżdżalnia i wjeżdżalnia.* Г. (4; 6, 25): *teraz to wszystko powwalam z powrotem. Powwalam a nie powywalam.* Г. (3; 7, 18): *postawię ciebie do kąta i poproszę tatusia i cię nie wystawi* Г. (3; 11, 29): *tu jest wpływ, a gdzie wypływ?* — место, через которое втекает и вытекает вода. М.: *bo pan kierownik może im pracę wymówić.* Г. (4; 6, 25): *a potem nową wmówić?* Г. (3; 7, 14): *ja się z tobą rozminąłem i zminąłem.* Мгновенное значение, выражаемое суффиксом *-na-*, выступает отчётливо в следующих примерах: Г. (3; 9, 14): *żeby było tylko trochę wody, to tylko raz trzeba pompnąć?* Г. (4; 0, 2): *bo cię łaskotnę zaraz.* Итеративные глаголы создаются при помощи суффиксов *-uje*, *-ywa-*, *-a-*, также как глаголы, развёрнутые при помощи префиксов: (4; 0, 1): *a nie będziesz mnie wiecej uduszała?* (4; 5, 2): *bo ja tak powoli wcelowuję*, В качестве дополнительного показателя итеративности выступают тематические параллелизмы: (4; 2, 15): *ja ci mogę tak codziennie ustampiać.* М.: *nie podchodź do tapczana, bo tapczan ciężki i niebezpieczny.* Г. (2; 2, 12): *a babci tapczan spieczny?* М.: *bardzo mnie to niepokoi.* Г. (3; 0, 26): *a mnie pokoi.* М.: *przez te chore paluszki to mamusia taka niezdara* Г. (2; 5, 15): *a przez te zdrowe to zdara?*

В некоторых из приведённых выше примеров слово, являющееся основой сравнения, не находится в непосредственной близости к неологизму, но его можно найти в более ранних высказываниях ребёнка или разговаривающего с ним взрослого. Неологизм является инструментом анализа результатов наблюдения внеязыковой действительности, но одновременно, (а иногда и прежде всего), реакцией на другое слово. В неологизмах этого типа с психологической точки зрения главная составляющая (основа) играет отождествляющую роль, а дифференциация происходит за счёт второстепенных составляющих.

Ситуация меняется, когда противопоставляются словообразовательные основы двух слов — в этом случае повторенный суффикс становится показателем подобия того, что связывает неологизм с названием, являющимся основой сравнения (и соответствующие им понятия). М. (Гжесю): *jakiś ty grubas*! Г. (2; 10, 0): *a ty chudas*. М.: *trzeba tak trzymać, żeby nie ruszały się tylne kółka*. Г. (2; 10, 21): *ale przodnych nie?* Г. (3; 2, 29): *a to co ciocia ma, jak się nazywa, kurteczka?* Т.: *nie, jesionka*. Г.: *a moja zimionka. I jesionka i zimionka*. Г. (3; 3, 13): *Jak byłem z ciocią, to widziałem dwie fabryki. Pewnie cegielnie, bo cegły leżały*. М.: *ale chyba tu w śródmieściu nie ma cegielni*. Г.: *no to kamielnie, bo leżały kamienie*. М. (Гжесю): *całą kuchnię tym lakierem zasmrodziłeś*. Г. (4; 5, 12): *a nie, bo zapachniłem*. *Это не синяк, а красняк*, Ч. 43; *корова не бодает, а рогает*, Ч. 43; *утки утьком идут, а гуси гуськом*, Ч. 43. Как легко заметить во многих из вышеприведённых примеров, при отделении второстепенных формантов часто наступает расширение за их счёт словообразовательной основы, перинтеграция, благодаря которой возникают расширенные суффиксы (*kamielnia*), а зачастую расширяются и словообразовательные основы: (*odetkać-zadetkać*).

Осуществляя противопоставление словообразовательных основ, одновременно ребёнок достигает (в суффиксальных формациях) созвучия конечных слогов, удовлетворяя таким образом своё влечение к рифмам.

(4) Пожалуй только этому влечению можно приписать осу-

ществление следующих словообразовательных трансформаций: Г. (2; 4, 0): *konik-polnik.* М.: *konik polny, synku.* Г.: *nie, konik-polnik.* Г. (2; 5, 4): *Agatka miotką zamiatka.* Г. (3; 3, 29): *samolocisko-odrzutowisko.* Г. (4; 3, 1): *mamico-wilkico.* Г. (4; 5, 21): *mamica-niedobrzyca.* Г. (4; 8, 11): *rękawiczki-kieszoniczki. Начало — кончало; ты глухой — я слухой; далёкое — близёкое;* Ч. 164. *Старинный — новинный,* Ж. (7; 6, 29) Гв. 333.

(5) Существует большая группа неологизмов, возникновение которых не обусловлено, как нам кажется, непосредственным противопоставлением значений или функций формантов, реакциями на слово. Непосредственный импульс к их созданию берёт начало во внеязыковой действительности, создающей потребность в названии, причём мы не можем установить, какие модели обусловили словообразовательную форму неологизма. Создаётся впечатление, как будто они появляются внезапно, без усилия со стороны ребёнка, как результат действия тенденции к сокращению высказывания, замене оборота, состоящего из нескольких слов, одним словом, как результат автоматизированного механизма преобразования слов. Интересен факт, что преобразованию подвергаются в первую очередь такие словообразовательные формации, в которых формант несёт на себе т. наз. грамматическую функцию (мутационную и транспозиционную), преобразует одни части речи в другие и устанавливает соотношения между лицами и предметами, а также их отношения к признакам и действиям. Таким путём создаются многочисленные отглагольные и образованные от прилагательных существительные, отглагольные и образованные от существительных прилагательные, образованные от существительных и прилагательных глаголы, и т. д. Словообразовательные форманты, являющиеся показателями грамматических связей, определяющие принадлежность формации к определённому классу понятий (классы частей речи, подлежащные структуры типа "*ten kto, to co*" в отличие от сказуемостных структур типа "*to że*") применяются ребёнком в практике для преобразования словообразовательных основ. Однако на этом уровне

развития мышления умственному анализу поддаются только смысловые функции морфем и только их можно сравнивать и противопоставлять в словах. Общеструктуральная функция суффиксов (их важнейшая словообразовательная функция) недоступна умственному анализу ребёнка. Это и понятно — мышление о категориальной (понятийной) функции словообразовательных формантов требует высокой степени абстракции, к которой ребёнок дошкольного возраста, располагающий управляемым ситуацией, конкретновоображающим мышлением, неспособен. Зато он способен практически оперировать грамматическими "генерализаторами", которые легко возникают благодаря гибкости нервной системы ребёнка и лёгкости применения всевозможных динамических шаблонов на этом этапе развития. Показанная выше на примерах лёгкость, с которой ребёнок копирует словообразовательную структуру отдельных слов делает понятным процесс создания этих грамматических генерализаторов. Впоследствии они реализуются в неологизмах, создаваемых спонтанно, как реакция на обусловленные ситуацией стимулы. Интересно то, что эта схема языковой реакции на новую ситуацию, задерживаемая по мере того, как ребёнок увеличивает запас слов и овладевает нормами языка, "воскресает" при сильном эмоциональном возбуждении. Я имела возможность проследить это на примере моего сына: немногочисленные неологизмы, созданные, когда ему было 6-7 лет, были результатом сознательной шутки (как это часто случается и в разговорной речи) или появились в ситуации сильного возбуждения, когда нехватало времени на размышление и поиски в памяти соответствующего средства выражения, а стремление к сокращению высказывания было максимальным. Г. (5; 0, 14): *roz ... tego, rozdziadkować ci orzechy?* (со смехом). Г. (6; 2, 5): *od ... tego, odpapieruj mi to* (громко смеётся). (6; 0, 7): *a, jest kolarówka! mnie się nie chciało powiedzieć czapka kolarska i mówię kolarówka.* (6; 3, 5) (очень взволнованный, противореча): *będzie mi tu wgadywała* '*wmawiała*'. (6; 3, 10): *aha, będę sobie spodnie rozkantowywał*! (возмущённый; я просила, чтобы он снял брюки со стула, который был мне нужен). (6; 5, 2): *jaki świetny śnieg! dobry do saniowania.* Описанные механизмы образования вновь созданных слов имели место у всех детей, даблюдаемых в течение длительных периодов времени. Обобщая созданные мною данные, можно констатировать, что с возрастом увеличивается количество противопоставлений (с 18 до 49,8%) за счёт таких формаций, в которых нельзя отыскать источник таких противопоставлений (с 68,9 до 38,7%).

ОЧЕРЁДНОСТЬ ПОЯВЛЕНИЯ СЛОВООБРАЗОВАТЕЛНЫХ КАТЕГОРИЙ

Сравнение польских и русских данных позволяет констатировать, что хотя темп возникновения очередных словообразовательных категорий в речи наблюдаемых детей был неодинаков, очерёдность "введения в употребление" этих категорий очень схожа. В соответствии с наблюдениями А.Н. Гвоздева в первую очередь выделяются экспрессивные уменьшительно-ласкательные и увеличительно-уничижительные показатели. Г. (2; 0, 2): *krypka* ('вид лодки', '*krypa*'), (2; 0, 12): *mewka*; услышав новое слово *skutek* Г. (2; 2, 15): *skutek mały, duży skut*; Г. (2; 2, 7): *duży pies, piesisko.* Безсуффиксальные формации в этой же роли у русского ребёнка, более поздние: *суч таленький* ('сучок') Ж. (3; 10, 6).

Позднее появляются аналогичные названия носителей признаков (атрибутивные существительные — nomina attributiva), предметов и исполнителей действия (названия действующих лиц — nomina agentis), отглагольные и образованные от существительных, а также существительные, называющие действия (имена отвлечённые действия и состояния — n. actionis i nomina acti) — последние несколько позднее у русских детей. Г. (2; 0, 11): *osobówka* ('auto osobowe' — 'легковая машина'), Г. (2; 2, 10): *zgrzytacz Grześ jest*, Г. (2; 2, 26): *pan śmieciarzysta* (sprzątający śmietniki — 'убирающий мусорные ящики'), Г. (2; 2, 21): *przypinacz do przypinania*, Г. (2; 2, 4):

tarka do tarczenia, Г. (2; 5, 25): *Grześ załatwił meldunkowanie* ('źameldowanie' — 'прописку'). *Вкусняшка* Ж; (2; 7, 8), *постелилка* Ж. (2; 4, 13), *курилка* Ж. (2; 7, 6), *перебиральщик* Ж. (3; 5, 4), *делальщик* Ж. (3; 11, 6), *ножник* Ж. (3; 5, 1), *продаванье* Ж. (3; 4, 22).

Приблизительно в это же время отмечено появление первых аналогичных глаголов и существительных, образованных от приветственных, звукоподражательных и др. оборотов, глаголы, образованные от глаголов (префиксация, а также замена суффиксов), существительных и прилагательных.

Г. (2; 0, 12): *nie cześciaj*! (от слова *cześć* -привет!), Я.П. (1; 11, 23): *sznurek od dzińdzinia* (звонка), Г. (2; 2, 10): *podgasić trzeba ten gaz, nie zgasić*, М.С. (2; 2, 14): *nie wrzeszczy tylko powrzeszczuje*, Г. (2; 1, 28): *zasoczyło tu się* ('сделались пятна от сока'), Г. (2; 3, 21): *Grześ się zaciemni, bo tam jest za ciemno*, Б. (3; 5): *zakluczyć*, *odkluczyć*. *Покойночиться*, Ч. 31; *исчистить*, Ж. (2; 5, 3); *молоточить*, Ж. (2; 1; 3); *раскрылился*, Ж. (2; 9, 11); *вытрнул нос*, Ж. (3; 4, 22); *я красавлюсь*, Ч. 15; *заключить*, Ч. 67.

В качестве первых аналогичных прилагательных появляются прилагательные, образованные от существительных: как польскце, так и русские дети создают их в большом количестве, часто заменяя ими конструкцю существительное спредлогом (ср. Гв. 450).

Г. (2; 1, 24): *świeczki sosenkowe* (·ёлочные'), Г. (2; 2, 10): *lampka prądowa* (от *prąd* — 'электричество'). *Чангина мама* Ж. (2; 8, 9); *коровный дом*, Ж. (3; 5, 2); *паровозный сон*, Ж. (3; 6, 1).

Прилагательные, образованные от прилагательных, вместе с аналогичными формами изменения по степеням сравнения, прилагательные отглагольные и иногда образованные от других словообразовательних основ, а также существительные, обозначающие признаки (имена отвлечённые качества — nomina essendi) появляются значительно позднее. Объясняется это значительно более поздним появлением названий признаков в речи ребёнка: Г. (3; 1, 24): *łyżeczka najmała*; М.С. (2; 3, 18): *dym okropliwy*; Г. (2; 4, 17): *brzęczna butelka*; Г. (2; 10, 6):

zakrętna ścieżka; Г. (3; 1, 23): *niebezpieczność. Сластота*, Ж. (4; 8, 5); *вот и нет ровности*, Ж. (6; 6, 10).

Также созданные по аналогии наречия и формы изменения наречий по степеням восходят к более позднему периоду развития словотворчества по принципу аналогии в речи ребёнка: Г. (2; 5, 3): *dużej mi nalej* ('больше'); та же форма у Ц.Б. (4; 4), Г. (2; 11, 10): *głębiutko* ('глубоко — уменьшительно'); Г. (2; 9, 26): *po zebrzemu*, Г. (3; 1, 1) *trzaskliwie. По хромому*, Ж. (6; 6, 18); *больновато*, Ж. (6; 8, 11); *по самоварему*, Ж. (4; 9, 9).

В польских и русских материалах наблюдается распространение категории изменения по степеням сравнения на существительные и наречия, не изменяющиеся по степеням: М.: *masa dzieci na sali*. Г. (4; 9, 11): *a dorosłych jeszcze masniejsza masa*. Г. (4; 5, 0): *ja jeszcze ledwiej idę niż ty. Решето — решетей*, Ж. (5; 5, 26); *дым — дымей*, Ж. (6; 5, 30); *тогда толкотней* ('больше толкотни'), Ж. (6; 10, 14); *леней*, Ж. (6; 6, 21). Анализированное нами явление словотворчества по аналогии обычно выступает в дошкольном возрасте, а нередко оно имеет место и позже. Дальнейшее его развитие заключается в создании аналогичных формаций перечисленных выше категорий при помощи всё новых словообразовательных показателей и неологизмов, относящихся к богатой группе существительных, образованных от существительных: существительные мужского рода от существительных женского рода, и наоборот, (названия лиц, животных, даже предметов), существительные единичные от собирательных, и наоборот, названия детёнышей, жителей определённых географических районов и аналогичные географические наименования, названия мест, предметов и т.д. Кроме того, создаются сложные существительные, прилагательные и даже глаголы, аналогичные числительные и местоимения. Количество этих более мелких семантических групп, а также число общих словообразовательных категорий, встречающихся в речи разных детей, конечно, неодинаково, однако очерёдность их появления довольно схожа. Стоит подчеркнуть выступающее неоднократно структуральное сходство, а иногда даже тождество неологизмов польских и русских детей, например: *zakluczyć — заключить, mycha — мыха* Ж. (4; 5, 8), *jeż-jeżyca, ёжик — ёжица* Ж. (4; 4, 27).

Приведённый анализ позволяет утверждать, что у основы аналогичного словотворчества польских и русских детей лежат тождественные психические механизмы, обусловленные закономерностями процесса психического развития ребёнка. Многочисленные совпадения в рассмотренных материалах являются кроме того результатом сходства польской и русской словообразовательных систем.

БИБЛИОГРАФИЯ

К. Чуковский, *От двух до пяти.* Изд. 11, Москва 1965
M. Dokulil, *Tvoření slov v češtině.* t. 1. Praha 1967
W. Doroszewski, *Kategorie słowotwórcze.* Warszawa 1949
Podstawy gramatyki polskiej. Wyd. II, Warszawa 1962
А.Н. Гвоздев, Вопросы изучения детской речи, Москва 1961
J. Piaget, *Mowa i myślenie dziecka*, Lwów 1929
Studia z psychologii dzieca, Waszawa 1966
S. Szuman, *Rozwój pytań dziecka*, Warszawa 1939
Rola działania w rozwoju umysłowym małego dziecka, Wrocław 1955 (red.) *O rozwoju języka i myślenia dziecka*, Warszawa 1968

DEVELOPMENT OF CHILDREN'S VOCABULARY

PŘEMYSL JANOTA

(Praha)

Investigations into the growth and development of children's vocabulary are not novel in Czech phonetic literature; the well-known monographs of Ohnesorg not only give an exhausting description of the phonetic development of two Czech-speaking children, but at the same time, complete vocabularies of the children. In another study (1955) Ohnesorg compares the two vocabularies and presents an analysis thereof from the lexical point of view; amongst other things he finds considerable overlapping between the vocabularies and offers a detailed explanation of the differences observed.

In order to arrive at a more general picture of the growth of children's vocabulary, a comparison of his lexical material with two other existing vocabularies has been attempted, the results of which will be given in the present paper. The comparison is thus based on the vocabularies of Ohnesorg's two children born in the years 1942 and 1944, on the vocabulary given in Pačesová's book *The Development of Vocabulary in the Child* (Brno, 1968) and on the record of the language development of my daughter Eva, born in 1958. No description of the phonetic development will be given here; in general her phonetic realizations were quite in accordance with Ohnesorg's findings and were in no respect exceptional. In all, a vocabulary of 921 words was used for the comparison with Ohnesorg's and Pačesová's data which amounted to 1920, 1613 and 500 words, respectively.

First of all, the amount of overlap of the lexicon in all four vocabularies was examined. In the test, the range of all vocabularies was limited to the first 500 words to obtain samples of identical size; in further treatments the whole number of lexical units was used. Proper names were disregarded in all procedures with the only exceptions of the child's own name, which was listed, however, as a single entry "own name".

In a simple matrix the numbers of overlapping words are given (Fig. 1).

	K	M	J	E
K	—	282	224	204
M	282	—	240	210
J	224	240	—	204
E	204	210	204	—
tot.	710	732	668	618

Fig. 1

It is not surprising that the number of words common to the boy K. and his sister M. is quite high. The amount of overlap is lower in other pairs, but the difference is smaller than was expected. Expressed in percentages (Fig. 2), the overlaps amount

	K	M	J	E
K	—	56%	45%	41%
M	56%	—	48%	42%
J	45%	48%	—	41%
E	41%	42%	41%	—

Fig. 2

to 56 % in the pair "brother and his two years younger sister M." and it is only 8 % lower in another pair, M. and J., "J" being a BOY, more than ten years younger, born in an other part of Czechoslovakia (in the city of Brno, the other children being born in Prague). Identical numbers in the intersections of K and E and J and E are incidental, showing clearly, nevertheless, that there is no noticeable difference in the cross-comparisons E-J vs.

E-K. The largest column total in the M-column suggests that the girl M. has the largest overlap with all the other children's vocabularies. Further, correlation coefficients per pair of children have been computed; they admit the conclusion of a significant association between the children K. and M., which means that the overlap of vocabularies in this pair can be regarded as implying a real correlation. Finally, the overlap in the vocabulary of all four children was computed: they have, in samples of the first 500 words each, the quite large number of 120 words (= 24 %) in common.

The matrices have been shown here, because the procedure of comparison described below does not readily lend itself to further statistical treatment. The simplest method of comparing the vocabularies is perhaps to state the number of vocabularies in which a particular word can be found. In such a case, an equal weight is attributed to each member of the corpus, irrespective of the serial number of its first occurrence. We may maintain, however, that the 'early' words, with the exception of words of an interjectional or onomatopoeic character (which are later replaced by other word forms) and apart from their imperfect phonetic realization, are in general more essential and characteristic for child language. An attempt has thus been made to divide all the words into ranks of various weight according to the order of their first occurrence, and to compare the ranks rather than the individual words.

When studying the growth of the child's vocabulary, one easily discovers that the growth of the vocabulary is roughly logarithmic, or — in other words — that a logarithmic scale gives a better approximation of the growth curve than a linear one. Although the curves have different starting points according to the age when the first word was uttered, they have similar slopes and run more or less parallel to each other, especially around the age of two years. This is in agreement with the currently common practice of treating 'the first 50, 150 etc. words' rather than the number of words at a certain age.

For the purpose of comparison, the groups of words can be acquired by selecting equidistant points on the logarithmic scale, which thus gives groups of an unequal number of members. The groups (ranks) can then be compared to those of other vocabularies.

In the present investigation, the vocabularies were divided into six ranks, labelled 6 to 0, corresponding roughly to three months' increments of vocabularies and including the 1st-3rd, 4th-10th, 11th-32nd, 33rd-100th, 101st-316th, 317th-1000th and 1001st-(3162nd) words, respectively. All four vocabularies were treated in this way and compared with one another. The position of each word was determined by the total of its four rank numbers, its 'score'. For the sake of convenience, the words have been ordered alphabetically within each rank.

The resulting vocabulary does not represent an 'average', 'normal', or 'standard' vocabulary, but a possible vocabulary, in which the less current words have been suppressed, but — at the same time — the weight of particular words, given by their order of occurrence, retained. A part of the vocabulary, giving the first 244 words, is presented in the Appendix. Deviations fiom this 'treated' list of words may perhaps — with careful interpretation — give some hints about the child, its educational environment, contacts, interests etc. It is perhaps not necessary to attempt to discuss the vocabulary word by word; permit me to make a few short remarks instead.

I mentioned at the beginning of my paper that only proper names were excluded from the vocabularies treated; other entries were retained, which leads to a high number of interjections, nursery words, onomatopoeic words and even to one or two words which may not be generally known to all children. Though I had more children's vocabularies at my disposal, I preferred to use only the above-mentioned ones as being more reliable and more carefully prepared.

The attempt to derive a word list from several existing vocabularies is meant to serve everyday practical requirements. I am thinking not only of the use in speech audiometry with small children, in which the choice of unfamiliar words may sometimes cause difficulties, but especially of the need for a suitable amount of material available for use in training children with speech and/or hearing handicaps. The careful selection of training material seems essential above all in the training and education of those children in whom the development of a complete vocabulary can never be expected.

REFERENCES

Ohnesorg, K.,
1948 *Fonetická studie o dětské řeči* (Praha).
1955 "Růst slovní zásoby v dětské řeči", *Universitas Carolina, Philologica* 1:1, 85-107.
1959 *Druhá fonetická studie o dětské řeči* (Brno).
Pačesová, J.,
1968 *The Development of Vocabulary in the Child* (Brno).

APPENDIX

Score:
19 *máma, maminka* 'mummy'
táta, tatínek 'daddy'
18 *haf, hafan* 'bow-wow!'
tutú, tú 'beep-beep!'
17 *auto* 'car'
bú (bučení) 'moo-moo!'
pa! papa! 'bye bye!'
16 *papat, spapat* 'to eat'
15 *babička* 'grandmother'
bác! 'bang!'
hajat 'to sleep'
vl.jm.dítěte 'own name'
ne, ne- 'no, not'
tam, tamhle 'there'
tik tak 'tick-tock'
14 *bota* 'shoe'
kytička 'flower'
teta 'auntie'
13 *am, ham* 'yum-yum'
ano 'yes'
bebé, bebí 'pain'
děda, dědeček 'grandfather'
tady, tadyhle 'here'
tak, takhle 'so'
12 *baf!* 'boo!'
dítě 'child'
hačat 'to sit'
hapat 'to fall, tumble'
hop! 'jump!'
hyja, hyje! 'gee up!'
kuk! 'peep-bo!'
11 *bé (bečení)* 'baa-baa!'
bumbat 'to drink'
hodný 'good, kind'
holčička, holka 'little girl'

hou, houpy 'to swing'
hú, hučet 'to hum'
jeti 'to go, ride'
medvěd, medvídek 'teddy-bear'
míti 'to have'
š š 'about a train'
10 *býti* 'to be'
čiči! 'pussy!'
jéje, jé! 'oh!'
ještě 'once more'
kabát 'coat'
kakat 'to defecate'
kolo 'wheel, bicycle'
malý,malinký 'small, tiny'
mňau, mňoukat 'to miaow'
pojď (sem)! 'come here!'
ryba 'fish'
slon 'elephant'
ucho 'ear'
voda 'water'
9 *aeroplán* 'plane'
běhat 'to run'
bum! 'bang!'
čaj 'tea'
čáp 'stork'
dát, dávat 'to give'
e (něco ošklivého) 'nasty'
elektrika 'tram'
hajs! 'hot!'
hlava 'head'
houba 'sponge'
houpat se 'to swing'
chléb 'bread'
jíti 'to go'
kachna 'duck'
kluk 'boy'
knoflík 'button'
kočička 'pussy-cat'
kohout 'cock'
kokokodák! 'cluck-cluck!'
králík 'rabbit'
lžíce 'spoon'
maso 'meat'
pršet 'to rain'
pusa 'mouth'
spát, spinkat 'to sleep'
strýc, strýček, strejda 'uncle'
šaty 'clothes'
tma 'dark(ness)'
ty, tyty 'you'
zub 'tooth'
8 *aha* 'aha'
au! 'ow! ouch!'
děkovat 'to thank'
díra, dírka 'hole'
dva 'two'
had 'snake'
hadr 'duster'
hanba 'shame'
hapčí, hepčí 'ashoo!'(sneeze)
hezký 'nice'
hodina, -y, -nky 'hour, clock, watch'
horký 'hot'
hřeben 'comb'
husa 'goose'
kalhoty 'trousers'
kartáč, kartáček 'brush'
kde 'where'
klíč 'key'
kniha 'book'

lopata, *lopatka* 'spade'
mé mé! *(mečení)* 'meh-meh!' (goat)
míč 'ball'
můj 'my'
nic 'nothing'
nos 'nose'
oko, *očko* 'eye'
paci! 'pat-a-cake!'
panenka 'doll'
pipi! 'chip-chip!' (chicken)
pryč 'away'
pták, *ptáček* 'bird, birdie'
ruka 'hand'
sám 'alone'
svítit 'shine'
tužka 'pencil'
utírat, *utřít* 'wipe'
vlasy, *vlásky* 'hair'
vstávat, *vstát* 'to get up'
zpívat 'to sing'
žába 'frog'
7 *aá* *(čurat)* 'to urinate'
bolet 'to ache'
brambora 'potato'
brejle 'glasses'
bříško 'tummy'
cililink! 'ting-a-ling!'
číst 'to read'
čistý 'clean'
čurat 'to urinate'
deka, *dečka* 'blanket'
dělat 'to do, make'
dobrý 'good'
dolů 'down'
držet 'to hold'
fff *(kouřit)* 'to smoke'
haló 'hallo'
hráti *(si)* 'to play'
hrnec, *hrneček* 'cup'
hruška 'pear'
chovat 'to nurse'
chtít 'to want'
já 'I'
káva, *kafe* 'coffee'
kočár, *kočárek* 'pram'
kostka 'block'
křičet 'to shout, cry'
kůň 'horse'
lézt 'to creep, crawl'
loď 'ship, boat'
noha 'foot, leg'
ocas 'tail'
okno 'window'
padat 'to fall'
pán '(gentle)man'
papání, *papú* 'food'
papír 'paper'
plenka 'nappy'
pozor! 'be careful!'
prosit 'to ask, beg'
psát 'to write'
rádio 'wireless, radio'
světlo 'light'
šátek 'scarf'
taky 'also'
taška 'bag'
utéci, *utíkat* 'to run away'
vana 'bath'
6 *a* 'and'
baba! *(hra)* 'children's game, tag'

balon 'ball, balloon'
bos 'barefooted'
buchta 'cake'
co 'what'
čepice 'cap'
daleko 'far away'
doma 'at home'
domů '(to go) home'
dost 'enough'
drak 'kite'
druhý 'second'
dveře 'door'
duc! 'bump!'
foukat 'to blow'
holub 'pigeon'
hopitó! 'jump!'
hůl 'stick'
chr-chr 'grunt' (pig)
jablko 'apple'
kapsa 'pocket'
kašlat 'to cough'
knedlík 'dumpling'
koukat 'to look'
koza 'goat'
kráva 'cow'
krk, *krček* 'neck'
kuře 'chicken'
kutululú 'to roll' (a ball)
mašina, *-nka* 'machine, toy-train'
mésti 'to sweep'
mléko 'milk'
moc, *mnoho* 'many, much'
mokrý 'wet'
motýl 'butterfly'
mydlit, *mýdlo* 'soap'
myš 'mouse'
mýti 'to wash'
někdo 'somebody'
nésti 'to carry'
oheň 'fire'
opice 'monkey'
ovce 'sheep'
palec 'thumb'
pálit 'to burn'
pásek 'belt'
pata 'heel'
pes 'dog'
píti 'to drink'
plakat 'to cry, weep'
plavat 'to swim'
rád 'glad'
rukavice 'glove'
sedět, *sednout* 'to sit (down)'
slepice 'hen'
spravit, *-ovat* 'to mend'
státi 'to stand'
špína, *špinavý* 'dirt(y)'
šťáva, *limonáda* 'drink'
talíř 'plate'
ten, *ta*, *to*, *tenhle* 'this'
tepláky 'track-suit'
těžký 'heavy'
vařit 'to cook'
veliký 'big'
vlak 'train'
zima 'cold, winter'
zlato, *zlatý* 'gold'
zlobit 'to be naughty'
zobat, *zobák* 'peck, beak'

DISCUSSION:

PAČESOVÁ: (1) What is the frequency of nursery words and interjections (especially those of onomatopoeic origin) in the child's wordstock in the first stages of his speech development?

(2) Is the phonemic repertory in interjections and nursery forms different (or limited) from the phonemic repertory in other word categories?

ANSWERS:

(1) Their frequency is considerably higher; cf. Appendix.

(2) This has not been investigated in the present study.

STELZIG: Zum Problem des aktiven Wortschatzes: Notwendigkeit einer kontinuierlichen Entwicklung und Festigung (normierbar nach den einzelnen Entwicklungsstufen, national und international) Wortschatz mit zentraler Funktion im Kommunikationsprozess.

DVONČOVÁ: Ich will die Stellungnahme des Herrn Stelzig unterstützen. In unseren Schulen (in Bratislava) macht man sehr oft Untersuchungen über die Geschicklichkeit des Schülers im Lesen. Sekundär kann man sie auch für die Untersuchungen über den Wortschatz des Schülers halten. Aber die Geschicklichkeit ist auch mit vielen anderen Phänomenen verbunden. Wir brauchen also auch spezielle Untersuchungen über den aktiven Wortschatz des Schülers im Sinne, wie es Dr. Stelzig vorgeschlagen hat.

GEBRAUCH DER EIGENSCHAFTSWÖRTER BEI HÖRGESCHÄDIGTEN KINDERN

NADĚŽDA JANOTOVÁ

(Praha)

Während meiner Arbeit auf dem Gebiet der Reedukation (Hörtraining) hörgeschädigter Kinder habe ich im Laufe der letzten Jahre nicht nur den Hörübungen, sondern vor allem der Entwicklung des Sprach- und Sprechvermögens eine erhöhte Aufmerksamkeit gewidmet.

Bei dieser Arbeit habe ich öfters beobachtet, daß die Mängel in der Sprache der Schwerhörigen sich besonders beim Gebrauch der Eigenschaftswörter bemerkbar machen. Dabei hat sich nicht nur eine beschränkte passive Kenntnis mancher wenig geläufiger Eigenschaftswörter gezeigt, sondern auch eine auffallend kleine Anzahl der in freier Erzählung verwendeten Eigenschaftswörter.

Deshalb habe ich einige Untersuchungen durchgeführt, deren Ziel doppelt war:

(1) festzustellen, ob diese Beschränkung im Gebrauch objektiv vorhanden ist und sich experimentell feststellen läßt,
(2) im positiven Fall, eine Anzahl geeigneter Übungen vorzubereiten, um zu einer erwünschten Erhöhung im Niveau des Adjektivgebrauches zu gelangen, denn bei Adjektiven handelt es sich vor allem um deren näher bestimmende, erläuternde Funktion, die in manchen Fällen zur Deutlichkeit und Genauigkeit der Aussage beträchtlich beitragen kann.

Dies ist übrigens in voller Übereinstimmung mit dem Unterrichtsplan für den muttersprachlichen Unterricht, besonders für die vierte Klasse, in der die Kinder ziemlich ausführlich mit den Eigenschaftswörtern bekannt gemacht werden, und für die fünfte Klasse, in der man gründliche Übungen zur klaren und passenden Ausdruckweise anstrebt und nebst der geeigneten Wahl von Zeitwörtern auch die Eigenschaftswörter hervorhebt.

In derselben Weise gehen auch die Verfasser methodischer Handbücher vor (Ballay, Janáček, Hauser usw.). Ich verzichte auf eine nähere Zusammenfassung der betreffenden Literatur, denn diese würde den Rahmen dieses Referates überschreiten.

In der Vorbereitungsphase meiner Untersuchung habe ich die in dem Frequenzwörterbuch der tschechischen Sprache angeführten Daten mit denen aus anderen Quellen verglichen, z.B. mit den Wortschatzuntersuchungen, die hauptsächlich unter der Leitung von Prof. Příhoda ausgeführt worden sind. Außerdem habe ich — da die obengenannten Studien aus etwas älterer Zeit stammen — einige eigene Wortzählungen unternommen und ziemlich große Unterschiede im Vergleich mit meinem Material aus den gegenwärtigen Lese- und Lehrbüchern gefunden. Ich habe eine Gruppe von insgesamt 87 000 Wörtern bearbeitet, so daß die Zahlen für meine Orientation ausreichend sind.

Ich möchte nun ganz kurz einzelne Teste beschreiben, die ich den Hörgeschädigten und zwecks des Vergleichs auch normalhörenden Kindern vorgelegt habe:

(1) Eine Nacherzählung eines für diese Zwecke vorbereiteten Märchens, d.h. einer in der Art moderner Märchen aufgefaßten Erzählung (342 Wörter lang), jedoch eine sehr große Anzahl von Eigenschaftswörtern enthaltend. Das Märchen wurde:
   (a) von schwerhörigen Kindern schriftlich ausgearbeitet,
   (b) von normalhörenden Kindern schriftlich reproduziert,
   (c) von schwerhörigen Kindern mündlich nacherzählt und auf Tonband aufgenommen; zugleich mit dieser Aufnahme habe ich die Kinder einige Minuten ebenfalls auf Tonband über ihre Ferien erzählen lassen (weiter als 'freie Erzählung' bezeichnet).

Hauptsächlich interessiert uns natürlich die relative Zahl der verwendeten Eigenschaftswörter. Diese werden auf dem ersten Bild dargestellt (Abb. 1).

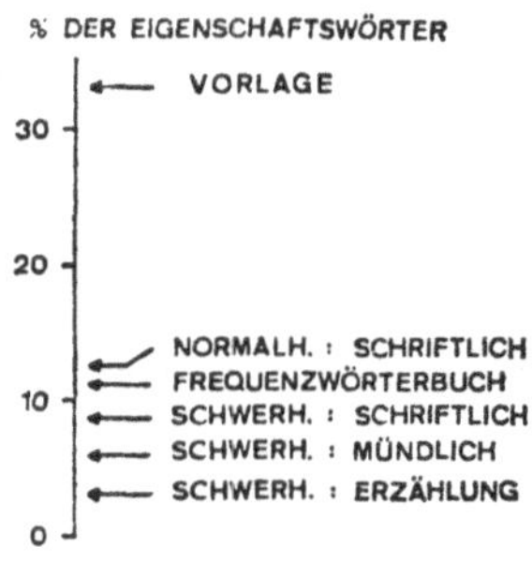

Es ist deutlich zu sehen, daß die Anzahl der Eigenschaftswörter, die in der Vorlage außerordentlich hoch war, in der Reproduktion stark herabgesetzt wurde, und zwar bis auf 6 % in der Nacherzählung auf dem Tonband. Außerdem ist hier noch eine niedrigere Prozentzahl dargestellt, die der 'freien Erzählung' entspricht. Dies ist allerdings nicht überraschend, da hier die Kinder durch keine Vorlage beeinflußt wurden.

In den Angaben übei die schriftliche Reproduktion des Märchens ist ein deutlicher Unterschied zwischen den schwerhörigen und normalhörenden Kindern bemerkbar. Ich möchte unterstreichen, daß die Prozentzahlen hier das Verhältnis zur gesamten Anzahl der verwendeten Wörter ausdrücken, d.h. von allen Wörtern haben die Kinder einmal 12,6 % und einmal 8,7 % Eigenschaftswörter benutzt. Wenn wir jedoch berechnen, um wieviel Prozent die Anzahl der Eigenschaftswörter bei Schwerhörigen kleiner ist als bei Normalhörenden, stellen wir fest, daß diese Anzahl um 31 % kleiner ist. Es handelt sich also um einen sehr großen Unterschied, der überzeugend sein mag (Abb. 2).

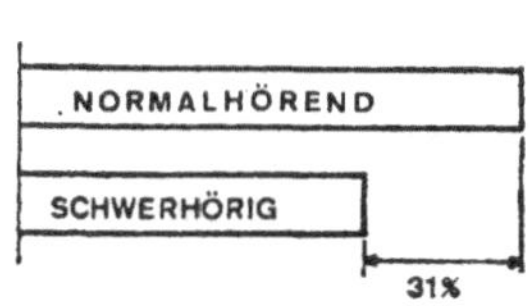

Weiterhin können wir versuchen, die Relation zwischen den in der Reproduktion vorkommenden Eigenschaftswörtern auf die Anzahl der Eigenschaftswörter in der Vorlage zu beziehen. Wir legen also die Anzahl der Eigenschaftswörter im vorgelesenen Märchen gleich 100 % und drücken in der folgenden graphischen Darstellung (Abb. 3) die Prozentzahl für die verschiedenen Typen der Reproduktion aus: "schriftliche Reproduktion — Normalhörende", "schriftliche Reproduktion — Schwerhörige" und "freie Erzählung — Schwerhörige".

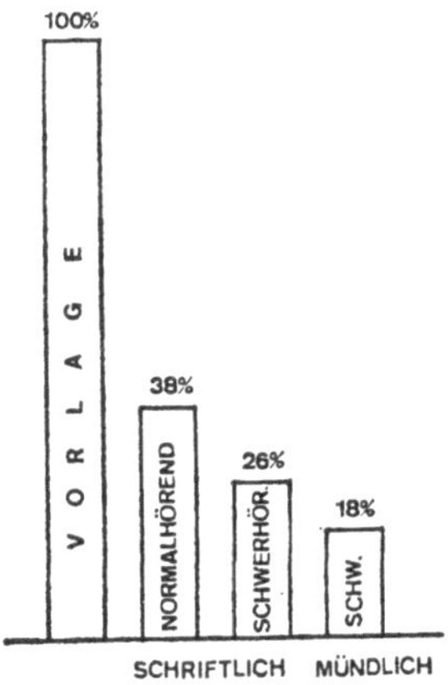

Ich möchte nicht zu viel Zeit durch weitere Erläuterungen zu diesem Test verlieren, vielleicht darf ich noch beifügen, daß für alle Nacherzählungen der Wiederholungsindex berechnet wurde, dessen Werte zwischen 1,09 und 2,47 schwanken:

	Wiederholungs-index
Vorlage	1,09
Schriftliche Reproduktion — Norm. . . .	1,20
Schriftl. Reproduktion — Schwerh. . . .	1,20
Mündliche Reproduktion — Schwerh. . .	2,47

(Den Wiederholungsindex berechnet man als Verhältnis aller Eigenschaftswörter zu den verwendeten verschiedenen Eigenschaftswörtern; er ist also gleich eins bei keiner einzigen Wiederholung.)

(2) Ziemlich ähnliche Resultate liefert auch eine Analyse der Stilaufgaben von schwerhörigen und normalhörenden Kindern. Im ganzen habe ich über 16.000 Wörter bearbeitet und möchte an dieser Stelle nur eine einzige Zahl nennen: die Anzahl der Eigenschaftswörter ist hier auch deutlich niedriger bei den Schwerhörigen, und zwar um 20 %.

(3) In der dritten Etappe meiner Untersuchung beschäftige ich mich jetzt mit der qualitativen Auswertung des besprochenen Materials; außerdem habe ich den Schwerhörigen und auch den Normalhörenden einen Test vorgelegt, der speziell auf das Gebrauch der Eigenschaftswörter gezielt ist. Diese Arbeit ist noch nicht abgeschlossen; die bisherigen Ergebnisse deuten an, daß den schwerhörigen Kindern die Bedeutung mancher Eigenschaftswörter nicht gut bekannt ist; dies ist besonders bei den nicht abgeleiteten Adjektiven der Fall.

Zum Schluß möchte ich zusammenfassen: In meinem Beitrag habe ich zu zeigen versucht, daß in der Sprache der schwerhörigen Kinder eine deutliche Verminderung im Gebrauch der Eigenschaftswörter vorliegt, die — meiner Meinung nach — durch eine erhöhte Aufmerksamkeit bei den Sprech- und Sprachübungen beseitigt oder verbessert werden kann.

DISCUSSION:

SIERTSEMA: Asks if any differentiation has been made between children of different social backgrounds. Tells about research in Holland into the causes of backwardness in school children which showed that children of 'intellectuals' had a far greater vocabulary and understanding of language than say, children from labourers families. If such backward children are to catch up so as to have the same chances as the others FROM THE START of school life, this catching up has to begin at a VERY EARLY AGE and that is so far an unsolved problem.

SPRACHANBILDUNG BEI DER GEHÖRLOSEN BLINDEN HELENE KELLER

LEON KACZMAREK

(Lublin)

0. Obwohl sich das Problem der Sprachanbildung und der Sprachentwicklung bei Hörgeschädigten seit langem im Mittelpunkt des Interesses der Surdologopädie befindet, ist und bleibt es wohl noch weiterhin Gegenstand eingehender Studien. An dieses Problem anknüpfend, halte ich es für nützlich zu erforschen, in welcher Weise die überall bekannte Helene Keller (27.6.1880 - 1.6.1968) die Fähigkeit erlangte, sich sprachlich zu verständigen, sogar mit der Lautsprache. Es wird auch für die Praxis wichtig sein. In meiner Betrachtung stütze ich mich auf: (1) H. Kellers *Geschichte meines Lebens*, (2) Auszügen aus ihren Briefen an Freunde, sowie (3) Berichte von John Albert Macy, des Redaktors des Buches, die gestützt sind auf Niederschriften und Hinweisen von Anna Sullivan.

1. H. Keller, die schon im neunzehnten Lebensmonat das Sehvermögen und das Gehör verlor (1 ; 6), was schon die damals beträchtlichen Fortschritte im Sprechen ausschloss, beherrschte einige Weisen (im ganzen 5) der sprachlichen Verständigung, und zwar:

2.1. Das in die Hand Schreiben und das Lesen mit Hilfe des Fingeralphabets. Dieser Unterricht war der erste und grundlegendste Weg zur Sprachanbildung (i.e.S; Kaczmarek, 1970) bei H. Keller. Der Anfang der Arbeit mit dem Mädchen, das damal: fast 7 Jahre alt war (es fehlten nur noch 3 Monate), war folgenders die Lehrerin, Anna Sullivan, begann am Handteller des Mädchens, welches gerade mit einer Puppe spielte, langsam das Wort *doll* 'Puppe' zu fingern. Die Kleine, die für dieses Spielen mit den Fingern Interesse zeigte, fing an, die Lehrerin nachzuahmen. Nach einigen Versuchen gelang es ihr, dieses Wort richtig zu reproduzieren. In den vielen darauffolgenden Tagen lernte Helene, über 30 Wörter fingern. Aber die ganze Zeit hindurch, verstand sie nicht sich mit ihnen zu bedienen, weil sie einfach nicht wusste, dass jedes Ding seinen eigenen Namen hat. Trotzdem A. Sullivan fingerte in die Hand Helenas den ganzen Tag alles, was sie erlebten und was sie machten.

Der 5. April 1887 bildet für Helene einen Wendepunkt. Die mehr als einen Monat dauernde Arbeit und die Mühe der Lehrerin beginnen nun zu fruchten. Helene verstand plötzlich, dass jede Sache ihre Bezeichnung hat, sie lernte mit diesen Namen zu umgehen, mit einem Wort, sie stand an der Schwelle der Sprachbeherschung, d.h. der Kenntnis der Symbole und der Distribution der Bauprinzipien der Texte und der Fähigkeit, diese zu lesen. Auf welche Weise diese Offenbarung zustande kam, ist allgemein aus den Berichten von A. Sullivan bekannt (Keller, 1967).

Als Helene begriff, "dass jede Sache ihren Namen hat und dass man mit Hilfe des Fingeralphabets diese Namen anderen Menschen übermitteln kann", bemühte sich A. Sullivan "in ihr weiteres Interesse für diese Sachen zu wecken, deren Namen sie mit begreiflicher Freude fingern lernte". Niemals war der Sprachunterricht Ziel, die Sprache war immer ein Mittel Wünsche und Erlebnisse zu übermitteln und der Sprachunterricht ging immer Hand in Hand mit der Erwerbung des Wissens. Solches Vorgehen war richtig, denn die grösste Geläufigkeit im Sprechen oder Schreiben garantiert den Kindern keine Gewandtheit in der Ausdrucksweise, wenn sie in ihren Köpfen keinen klaren Inhalt besitzen sollten. Helene lernte die Sprache durch den Kontakt mit dem Leben. Ihre Hände tasteten alle Gegenstände und kontrolierten Bewegungen aller Menschen, die in ihrer Nähe waren. Als Endergebnis des kaum 8 monatigen Unterrichtes war eine geläufige Beherrschung der Sprache (i.e.S.), die die 7 Jahre und 3 Monate und einige Tage alte Helene erwerben konnte. Ihre Aussagen sind nunmehr korrekt, sie bilden schon zweiklassige Signale. Die Grammatik, d.h. die Lehre von der Sprache samt ihrer unbeholfenen Klassifizierung, Nomenklatur, Deklinations- und Koniugationsparadigmata spielte im Sprachunterricht keine Rolle.

2.2. Das Lesen der hervorgewölbten Texte und das Schreiben der normalen Texte. Das Fingeralphabet war nicht der einzige Weg in der Gestaltung und Entwicklung der Sprache (i.e.S.) bei Helene, es war auch nicht das ausschliessliche Mittel zur Verständigung mit der Umgebung. Nicht mindere Rolle spielte hier das Lesen und Schreiben. Der Lese-Unterricht begann dann, als Helene lernte etwas auf der Hand zu schreiben. Es war ungefähr 3 Monate nach dem Beginn des Unterrichtes. Beim Lesen normaler Bücher half ihr die Lehrerin und die mit gewölbten Druck las sie allein, indem sie den Zeige-Finger der linken Hand längs der Linien führte und den durchgelesenen Text mit der zweiten Hand fingerte. Man muss besonders betonen, dass Helene viel las und zwar mit grosser Freude und Hartnäckigkeit, unabhängig davon, ob sie auf einer Seite von zehn Wörter nur ein Wort oder nur ein jedes zweite verstand. Anfänglich ging es ihr nicht um Inhalt, sondern vor allem um Aufsuchen bekannter Wörter. Wörter, die ihr bekannt oder unbekannt waren, speicherten sich im Unterbewusstsein Helenas und später, als sie schrieb und sprach, kamen sie ihr ganz naturell hervor. Ein Buch vom Anfang bis zum Ende und mit vollem Verständnis hat sie viel später gelesen. Was das Schreiben betrifft, das Schreiben einfacher Buchstaben (in die Hand), begann Helene, die jetzt 6 Jahre, 11 Monate und 15 Tage alt war, ungefähr drei Monate nach dem Beginn des Unterrichts, aber, wie A. Sullivan berichtet, nur als Zerstreuung. Doch schon nach anderthalb Monaten notiert Sullivan, dass Helenchen schon ausgezeichnet mit dem Bleistift schreibt. Ununterbrochene und systematische Arbeit gibt schnell ihre erforderlichen Erfolge. Nach zwei Jahren Unterrichts schreibt Helene fliessend (8 ; 11,4), unbefangen und stilistisch korrekt. Interessant ist die angewandte Technik des Lesens und Schreibens. Wie bekannt, Sullivan übernahm sie, indem sie bestimmte Vereinfachungen und Vervollkommnungen einführte, von Dr. S. Howe. Kurzen Bericht von ihr bringt *Logopedia* 10 (B. und L. Kaczmarek, 1971).

2.3. Das Lesen und Schreiben der Braille-Texte (mit der Hand und der Maschine). Nach 5 Monaten vom Beginn des Sprachunterrichtes (i.e.S.) mit Hilfe des Fingers (31.7.1887), als Helene schon verstand, ihre Gedanken aufs Papier zu bringen, brachte ihr die Lehrerin das Braille-Alphabet bei. Helene lernte es gern, da sie bemerkte, dass sie allein lesen kann, d.h. sie kann das kontrollieren, was sie geschrieben hat. Aus diesem Grunde hielt sie das Brailler Alphabet für das bequemste System für Verständigung der gehörlosen Blinden. Aus dieser M?glichkeit, gedanklichen Kontakt anzuknüpfen, machte Helene, was man betonen muss, gerne und mit Freude Gebrauch.

Die Bekanntschaft mit dem Braille-Alphabet war für Helene in jeder Hinsicht sehr nützlich. Es ermöglichte ihr, mindestens in gewissem Grade, selbständige Studien; das Vorbereiten für den Unterricht nahm ihr gewöhnlich viel Zeit in Anspruch, weil man jedes Wort ihr in die Hand fingern musste. Ferner erlaubte es ihr, ohne Schwierigkeit, ziemlich oft in diesem Alphabet gedruckte Texte, Handbücher und schöne Literatur auszunutzen. Das Lesen, wie wir es wissen, liebte Helene sehr. Endlich ermöglichte es ihr, mit Blindem unmittelbare Korrespondenz zu führen und eine Verfertigung von Notizen, gewöhnlich auf der Brailler Schreibmaschine.

2.4. Das Pochen mit dem Morse-Alphabet in die Hand und der Empfang mit Gefühl. Im Jahre 1889 oder 1890 beherrschte Helene Keller das Morse-Alphabet. Es ist klar, dass es bei der Sprach anbildung (i.e.S.) keine Bedeutung hatte, weil sie in dieser Zeit schon gut mit der Sprache (i.e.S.) umgehen konnte. Helene konnte sich aber dank dieser Kenntnisse mit allen verständigen, die dieses Alphabet kannten. Sie gebrauchte es mit grosser Freude.

2.5. Das lebendige Wort (Sprechen und das Lesen vom Mund mit Hilfe der Hand). Drei Jahre nach Beherrschung des Finger-Alphabets, am 26. März 1890, begann Helene die Sprache (i.w.S; Kaczmarek, 1970) zu lernen; sie war damals 9 Jahre und 9 Monate alt. Sie begann mit dem Sprachunterricht (i.w.S.), nachdem sie die Sprache (i.e.S.) schon beherrscht hatte, d.h. sie verstand und konnte sehr geläufig Texte mit Hilfe des Finger-, Latein-, Braille- und Morse-Alphabet zu bauen. Jetzt ging es um das Erlernen der Artikulation (= des Sprechens), sowie um die Kunst des Lesens fremder Aussagen von den Lippen vermittels der Finger. Das Verständnis für dieses Ziels ermunterte sie zur Anstrengung und erleichterte ihr, die vorkommenden Schwierigkeiten zu überwinden. Insgesamt 11 Stunden erteilte ihr Sary Fuller, die Direktorin der Horatius-Mann-Schule. Die Arbeitsmethode sah folgendermassen aus: Die Lehrerin gleitete Helenas Hand an ihrem Gesicht, damit diese die Lage der Zunge und der Lippen beim Sprechen eines Lautes fühlen konnte. Helene eignete sich alsbald alle Laute an. Seitdem übte Helene, indem sie von A. Sullivan beraten wurde, Tag und Nacht — wie sie schrieb — die Bindungen der Laute auf tausend Weisen. Sie redete ihre Spielzeuge, Steine, Bäume, Vögel und Tiere mit Begeisterung an. Auch machte sie schnell Fortschritte. Schon nach einer Woche konnte sie ganze Sätze aussprechen. Auf der 5. Tagung der Amerikanischen Gesellschaft für Propagierung des Sprachunterrichtes (i.w.S.) für Blinde am 8. Juli 1896 hielt Helene eine längere Rede. Sie brachte es nach 6 jähriger Arbeit zustande.

Was die Aussprache und die Stimme Helenas betrifft, so erreichte sie keine vollen befriedigenden Erfolge, trotz der ständigen Hilfe seitens der Fachleute. Sie änderte z.B. von Tag zu Tag die Aussprache desselben Wortes, bildete oft die Vokalfarbe um, nicht immer konnte sie das Wort entsprechend richtig lesen, das sie vorher noch nicht aussprach, obwohl sie es mehrmals geschrieben hatte. Ähnlich war es mit dem Lesen von den Lippen. Oft musste A. Sullivan, wenn jemand von ihrem Freundeskreis umsonst versuchte, sich mit Helene zu verständigen, zur Hilfe kommen und ihr die verlorenen Wörter in die Hand fingern. Kein Wunder, weil das Lesen des Kommunikats mit Fingern von den Lippen nicht recht zu handhaben ist. Wenn man nun global Helenas Erfolge beurteilen sollte, so ist es schwierig, A.G. Bells Meinung nicht zu teilen, wenn er sagte, dass die Fortschritte und die intellektuelle Entwicklung Helenas "ein eigenartiges Phänomen in der Geschichte der Lehre für Blinde ist".

3.0. Der Beweggrund dieses ausserordentlichen Erfolges waren drei Umstände sowie ein für damalige Zeit bahnbrechender, heute allgemein anerkannter (*Listy...*, 1964) surdologopädischer Grundsatz und die daraus erfolgenden Normen einer konkreten Tätigkeit.

3.1. Umstände: Helenas Persönlichkeit. Zweifellos war Helene ungewöhnlich begabt, aber auch sehr ehrgeizig. "Irgendwelche übermächtige Kraft — charakterisiert sich Helene — mächtiger sogar als die Sprache des eigenen Herzens hat mir befohlen, mit denen, die sehen und hören, zu rivalisieren." Helene zeichnete sich ebenfalls durch ausserordentlichen Fleiss und Beharrlichkeit aus.

Das pädagogische Talent von A. Sullivan. Die Lehrerin von Helene, A. Sullivan, hatte das Perkinsinstitut absolviert. Zur Arbeit an Helenas Sprachanbildung halfen ihr Erfahrungen, die sie im Institut von Dr. S.G. Howe erwarb, aber Autorin der Methode, die sie anwandte, war sie selbst. A. Sullivan verlor fast völlig als Kind das Sehenvermögen, das sie später teilweise wiedergewann. Ihre eigene Krankheit erlaubte ihr, die Psychik der gehörlosen Blinden Helene kennenzulernen. Dieser Umstand, d.h. gründliche Kenntnis der Fachpsychologie, verbunden mit einem aussergewöhnlichen pädagogischen Talent, das ständige Suchen nach neuen didaktischen und erzieherischen Wesen, der unbeugsame Wille, Entschiedenheit, Beharrlichkeit, Geduld — erlaubte es A. Sullivan, Helene aus dem Getto der durch das Schicksal benachteiligten herauszuführen und sie in die Welt der normalen Gesellschaft einzuführen.

Die Arbeitsmethode mit Helene. Es war, wie wir nach M. Grzegorzewska (1964) sagen können, die Arbeitsmilieux-Methode. Als Hauptproblem dieser Methode, um das sich alle anderen Probleme gruppierten, waren die Angelegenheiten des Alltags. A. Sullivan lehrte Helene frei und natürlich während des ganzen Tages, ohne sie durch 2 Jahre hindurch einem regulären Rigor systematischer Lehre zu unterziehen.

3.2. Der surdologopädische Grundsatz. Ich denke hier an den Umgang mit der gehörlosen Blinden wie mit einem hörenden und sehenden Kind. A. Sullivan zog daraus einzelne Normen, die sich weitgehend mit den heute empfehlenden logopädischen Geboten decken (Kaczmarek, 1970). Es sind:

(1) Unaufhörliches 'Sprechen' zum Kind beim Spiel oder bei der Arbeit über alles, was in seinem Umkreis geschieht und gemacht wird und immer mit vollen Sätzen.

(2) 'Sprechen' mit dem Kinde immer über Dinge, die es interessieren, und meiden von Gesprächen, die es langweilen oder die es nicht liebt.

(3) "Nichtberuhigung des Kindes, wenn es Fragen stellt, sondern möglichst genaue Beantwortung auf diese geben."

(4) In erster Linie die Akzentlegung, wie das heute die sowjetische Schule betont (Zykov, 1961), auf die Beherschung der Sprache (i.e.S.) auf dem einfachsten Wege, und dann erst soll das Sprechen und das Ablesen vom Mund erfolgen.

Richtigkeit der Behandlung im Sprachanbildungsprozess die gehörlose Blinde Helene als ein normalhörendes Kind zu betrachten, d.h. sie mit ganzen Aussagen zu bombardieren, bewiesen die erreichten Erfolge. Aus dem Hand-Hand-Wege erhaltenen Aussagen abstrahierte Helene Sprachsymbole und die Struktur der grammatischen Grundsätze und speicherte sie in ihrem Unterbewusstsein. Dank deren Kenntnis begann sie zu verstehen, was man ihr übermittelte, und bald darauf lernte sie Signale derselben Kategorie und in derselben Reihe zu bauen, wie ein normal hörendes Kind. Wir vermerken also bei Helene zuerst einklassige Signale (nackte Symbole, die gegenwärtige und nahe Sachen betreffen, und die erst in ihrer Konsituation verständlich sind = Wörter), dann zweiklassige (die aus den Symbolen auf Grund grammatischer Grundsätze gebaut werden und die unsichtbare Erscheinungen betreffen, die vom Absender und Empfänger in der Zeit und im Raum entfernt sind = Sätze). Aus verständlichen Gründen fehlen hier motivierte semantische Signale (onomatopeische Schöpfungen) sowie im strengen Sinne des Wortes Signale-Appelle (= Ausrufe der Art wie *hy*, *hm*, *my-my-my*). Die Funktionen der letzten, die Aufmerksamkeit der Umgebung hervorzurufen, erfüllt gewöhnliches Anklopfen auf der Hand des Mitsprechers.

4. Die erzieherische Arbeit A. Sullivans war, wie aus diesen selbstverständlich kurzen Ausführung zu ersehen ist, durchaus modern. Leider, ausser einer Bewunderung, die ihr zuteil wurde, zog die logopädische Praxis aus ihrer Tätigkeit grundsätzlich keine entsprechenden Schlussfolgerungen. Somit denke ich, es wird zutreffend sein, wenn ich in Anknüpfung an die erreichten Ergebnisse einige für die Surdologopädie wichtige Postulate darlege. Also:

(1) Die Sprachanbildung bei gehörlosen Kindern (darunter auch bei gehörlosen Blinden) soll man, wie das schon der polnische Forscher der 1. Hälfte des 19. Jahrhunderts, Jan Siestrzyński, betonte (Kaczmarek, 1965), so früh wie möglich in Angriff nehmen.

(2) Die Revalidierungsarbeit soll geführt werden: zunächst im Elternhaus, dann im Kindergarten für normalhörende Kinder (beide entsprechend vorbereitet) sowie im engen Kontakt mit ihnen mitarbeitende Surdologopäden.

(3) Die Grundlage für die Revalidieıungsarbeit sollte die Arbeitsmilieux-Methode sein in dem oben beschriebenen Sinne und Reichbreite.

(4) Die Sprachanbildung (i.w.S.) sollte auf einem dem Kinde am meisten zugänglichen Wege eingeleitet werden: mit Hilfe des Finger-, Normal- (lateinischen, kyrillischen), Braille-, Morse-Alphabets.

(5) Zum Festigen der gestalteten Sprache (i.e.S.) und ihrer weiterer Entwicklung sollen die Erzieher das Kind zu systematischem Lesen angewöhnen.

(6) Als Ergebnis all dieser Eingriffe soll das normal begabte Kind mit Gehörstörungen in solchem Grade die Sprache (i.w.S.) beherrschen, dass es eine entsprechend vorbereitete Schule für normalhörende besuchen konnte. Kinder, die das nicht zu erreichen im Stande wären (ich denke hier an Geistigbehinderte), bliebe nur eine Sonderschule übrig.

LITERATUR

(a) *Quelle*

Keller, H.,
1967 *Historia mojego życia* [Geschichte meines Lebens], übersetzt von Janina Sujkowska (Warszawa, Czytelnik).

(b) *Bearbeitungen*

Grzegorzewska, M.,
1964 *"Analiza wartości rewalidacyjnych metody 'ośrodków pracy'"* [Analyse der Revalidierungswerte der Arbeitsmilieux-Methode], *Wybór pism* (Warszawa, Państwowe Wydawnictwo Naukowe).

Kaczmarek, B.,
1971 *"Relacja Y. Dickensa o metodach kształtowania języka u głuchociemnych w Instytucie Perkinsa"* [Bericht von K. Dickens über die Sprachanbildungsmethoden bei der gehörlosen Blinden im Perkinsinstitut], *Logopedia* 10.

Kaczmarek, L.,
1965 *"Jan Siestrzyński, ein polnischer Vorläufer der Phonetik in der ersten Hälfte des 19. Jahrhunderts"*, *Zeitschrift für Phonetik, Sprachwissenschaft und Kommunikationsforschung* 18:4-5.
1971 *"Kształtowanie mowy u głuchociemnej Heleny Keller"* [Sprachanbildung bei der gehörlosen Blinden Helene Keller], *Logopedia* 10.
1970 *"Nasze dziecko uczy się mowy"* [Unser Kind lernt die Sprache] (Lublin, Wydawnictwo Lubelskie).

Listy o wychowaniu dziecka głuchego: *Poradnik dla rodziców*
1964 [Briefe über die Erziehung des gehörlosen Kleinkindes: Ein Lehrgang für die Eltern], mit Vorwort von Leon Kaczmarek (Polski Związek Głuchych, Wydawnictwo: Świat Głuchych).

Зыков, С. А.,
1961 *Обучение глухих детей языку по приншпу формирования речевого обучения*. Москва, Издательстьство Академии Педагогигеских Наук РСФСР.

DISCUSSION:

Borel-Maisonny: L'expérience d'Hélène Keller prouve que l'apprentissage d'une langue est avant tout celui de la pensée claire. Si les moyens habituels d'apprentissage de la langue par une méthode orale se heurtent à de trop grandes difficultés il faut recourir à d'autres formes même gestuelles, l'essentiel étant l'accès à la pensée et à la communication.

PHONETIK DER KINDER MIT VERZÖGERTER SPRACHENTWICKLUNG

JÓZEF T. KANIA*

(Lublin)

In der Entwicklung der Kindersprache kann man eine Reihe von Regelmässigkeiten beobachten. Ohne Rücksicht auf die Struktur der verwendeten Sprache lernen alle Kinder die Sprache auf dieselbe Weise. Ähnlich ist auch in allen Kulturen das Alter, in welchem die Kinder Grundlagen der Sprache beherrschen, obwohl man in primitiven Kulturen gewisse Verzögerungen bemerken kann.[1] Das Tempo der Aneignung der Muttersprache ist bei einzelnen Kindern verschieden.

Meistens erscheinen die ersten Wörter zwischen dem 1. und 2. Lebensjahr, die Entwicklung in dieser Hinsicht kann aber manchmal sogar zwei Jahre verspätet werden. Abgesehen von der Dynamik der Entwicklung ist jedoch der Sprachunterricht immer ein extensiver Prozess (er dauert einige Jahre). Es geschieht so, weil das Kind sich zum ersten Mal ein so kompliziertes Zeichensystem aneignet, das die Sprache bildet, und weil gleichzeitig mit der Entwicklung der Sprache auch andere physische und psychische Funktionen gestaltet werden, so z.B. die Sphäre der Bewegungen (Motorik), das Hören, Gedächtnis, Denken u.a.

In den Fällen der s.g. einfachen Verzögerung der Sprachentwicklung, die uns in diesem Referat vor allem interessiert, erscheinen die ersten Wörter am Ende des 2. oder sogar des 3.

* Not present in person.

1 Vgl. E. B. Hurlock, *Rozwój dziecka* [Entwicklung des Kindes] (Warszawa, 1960), S. 265.

Lebensjahres, oder die Kinder gebrauchen bis zu dieser Zeit lediglich einige Wörter mit einfacher phonetischer Struktur. Erst nach der Periode relativer Stockung erfolgt meistens bestimmte Beschleunigung der Entwicklung.

Dynamik der Sprachentwicklung ist also völlig verschieden im Vergleich mit der normalen. Folgende Frage fordert Erforschung: ist das Bild der Phonetik bei der verzögerten Sprachentwicklung mit dem allgemeinen Entwicklungsmuster einig oder besitzt es eigenartige Kennzeichen, die durch verschiedene Dynamik und verschiedene Entwicklungsstufe einzelner physischer und psychischer Funktionen bedingt werden. Dieses Problem möchte ich eben untersuchen, indem ich mich auf die Beobachtungen einiger polnischer Kinder mit verzögerter Sprachentwicklung stützen möchte.

Das meist auffallende Kennzeichen in der Phonetik dieser Kinder ist die Begrenzung des Lautinventars und verschiedenartige Verunstaltungen der phonetischen Struktur der Wörter. In äussersten Fällen kann die Begrenzung des Lautinventars so gross sein, dass der Konsonant *t* und Vokal *a* sogar im 50 % die vom Kinde produzierten Texte decken und im Bereich der phonetischen Struktur der Wörter beobachtet man nur die nuklearen Silben (vom Typ CV — Konsonant + Vokal) und charakteristische Schwierigkeiten bei der Einführung des Auslautkonsonanten in der Reihe CVC — *ko* ('kot' Katze), *pa* ('pan' Herr) u.a. Die Laute, deren Aussprache von den Kindern noch nicht beherrscht worden ist, werden in bestimmten Positionen substituiert (sie unterliegen der Substitution) oder infolge der Schwierigkeit im Wiedergeben der Wortstruktur weggelassen. Zu äussersten Ausnahmen gehört völliges Auslassen eines Lautes in allen phonetischen Positionen.

Im Bereich der Vokale werden am häufigsten die Mittelvokale *e* und *o* verwechselt.[2] Sie werden durch den Vokal *a* ersetzt. Verhältnismässig oft unterliegt der Substitution nur der Vokal *e*, bei dem richtig behaltenen *o*. Der Autor kam mit einer merk-

[2] Ich übergehe die nasalen Vokale, da sie in der modernen polnischen Sprache einen besonderen Status besitzen.

würdigen Aussprache bei einem Jungen im Alter 5 ; 10,11 in Berührung, der 3 Vokale: *a*, *u*, *o* produzierte und die übrigen durch *a* und *u* ersetzte (*e* > *a*; *i*, *y* > *u*).

Bei den Konsonanten bemerkt man mehr verschiedenartige Substitutionen. Stimmhafte Verschlusslaute, Spiranten und Affrikaten werden in der Regel durch stimmlose ersetzt, es können auch auf diesem Gebiet mehr oder weniger deutliche Schwankungen vorkommen. Bis zum Alter 5-6 kann auch begrenztes Inventar der weichen Konsonanten vorkommen. Zu weiteren Substitutionen gehören Ersetzungen der velaren *k*, *g* durch *t*, *d*, der Affrikaten und Spiranten *s z c ʒ*, *š ž č ǯ* und *ś ź ć ʒ́* durch *t*, *d* (und in den Fällen der 'stimmlosen Sprache' nur durch *t*); in anderen Fällen erfolgt Gleichschaltung der Realisation der 3 Reihen in der Gestalt von *s z c ʒ* oder *ś ź ć ʒ́*. Weitere Substitutionen sind: Ersetzung *x* durch *f* oder *k* (es liess sich nicht Ersetzung *x* durch *s* beobachten), *r* durch *l*, bzw. *j*, *l* durch *j*. Ausserdem soll man Schwierigkeiten in der Abgrenzung des *u̯* (als unsilbigen Laut) und *v* (bei der 'stimmlosen Sprache' *f*), Ausnahmeaussprache von *r* und *l* als interdentales *d*, *ø* (Nullaut) statt *x* im Anlaut usw. zur Kenntnis nehmen.

Besondere Schwierigkeiten, wie gesagt, beobachtet man bei diesen Kindern im Prozess der Aneignung der phonetischen Wortstruktur. Die Erscheinungen, wie progressive und regressive Fernassimilation, Methathese, Verminderung der Silbenzahl im Wort und Reduzierung der Konsonantengruppen halten sich hartnäckig eine längere Zeitperiode fest. Die polnische Sprache enthaltet eine grosse Anzahl komplizierter Konsonantengruppen — man kann also genau ihre Umgestaltungsprozesse untersuchen.

Am frühesten erscheinen die Konsonantengruppen im Inlaut. Es scheint auch Kennzeichen normaler Sprachentwicklung zu sein. In manchen Fällen verzögerter Sprachentwicklung kann man genaue Unterschiede im Verhalten derselben Konsonantengruppen im Anlaut und Inlaut feststellen. So bemerkte ich z.B. bei einem 6-jährigen Jungen:

pat (*ptak* 'Vogel')	aber: *apteka* ('Apotheke')
tuu̯ (*stol* 'Tisch')	aber: *postój* ('Haltestelle').

Die festgestellten Regelmässigkeiten sind von einer grossen Bedeutung für die logopädische Praxis — denn sie zeigen Positionen von welchen die Reedukation beim Mangel an Konsonantengruppen beginnen soll.

Ausserdem möchte ich andere interessante Erscheinung besprechen, die ich als STRUKTURALISIERUNG DER KONSONANTENGRUPPEN genannt habe. Unter diesem Begriff verstehe ich partielle oder völlige Gleichschaltung verschiedenartiger Konsonantengruppen, was die Universaliesierung eines Gruppentypus als Folge hat.

Eines der untersuchten Kinder gibt unbekannte Konsonantengruppen im Inlaut und Auslaut mit Hilfe von *-ńć-* // *-ńʒ́-* wieder, im Anlaut, wo in polnischer Sprache weder *n* noch *ń* irgendeine Konsonantengruppe anfangen können, kann dieser Typ von Strukturalisierung nicht erfolgen — deshalb wieder die Gruppen vereinfacht. Beispiele:

końćuu̯ek	(*kościółek* 'Kirchlein'),
guńć	(*gwóźdź* 'Nagel'),
gońʒ́e	(*gwoździe* 'Nagel'),
napańʒ́ik	(*naparstek* 'Fingerhut'),
ľińć	(*list* 'Brief'),
ǵjańʒ́i	(*gwiazdy* 'Sterne'),
tańći	(*czarny* 'schwarz'),
ḿeńće	(*miejsce* 'Platz'),
soneńʒ́ik	(*słonecznik* 'Sonnenblume'),
końć	(*kość* 'Knochen').

Strukturalisierung der Konsonantengruppen ist mit der Periode der Akzeleration der Sprachentwicklung verbunden. Es ist eine spezifische Art die Schwierigkeiten zu neutralisieren, wobei die Wörter relativ genau wiedergeben werden können. In beiden Fällen, wo eine deutliche Tendenz zur Gleichschaltung der Konsonantengruppen festgestellt wurde, konnte man gleichzeitig einen reichen Wortschatz bemerken. Man soll feststellen, indem man die Frage — ob die Strukturalisierung der Konsonantengruppen Ursache oder Folge dieses Zustandes ist — zur Zeit abseits lässt, dass sie als phonetische Erscheinung mit bestimmter Dynamik der Sprachentwicklung zu verbinden ist.

In einfacher Verzögerung der Sprachentwicklung können neben grundsätzlichen Konvergenzen mit normaler Sprachentwicklung auch abgesonderte Erscheinungen vorkommen. Was die Reihenfolge der Aneignung einzelner Laute anbelangt, kann man gewisse Abweichungen von grundsätzlichen Feststellungen von R. Jakobson[3] nachweisen, z.B. im Bereich der Vokale Anwesenheit von *o* und Mangel an *e*, Anwesenheit von *u* und Mangel an *i*.

Früheres Erscheinen der Laute *ś ź ć ʒ́* im normalen Prozess der Sprachentwicklung ist physiologisch bedingt — und zwar durch globale Arbeit der Zunge. Der Leistungssteigerung der Sprachorgane folgt die Abgrenzung der Laute *ś ź ć ʒ́* und dentaler *s z c ʒ*, erst später erscheinen alveolare *š ž č ǯ*. Bei vezögerter Sprachentwicklung, wenn der Prozess der Aneignung der Artikulation nicht parallel mit der Entwicklung der Bewegungsfähigkeit der Sprachorgane verläuft, erscheinen die Laute *ś ź ć ʒ́* viel später, nach der dentalen Reihe.

Von der verschiedenen Dynamik der Sprachentwicklung sind auch gewisse charakteristische Erscheinungen auf dem Gebiet der phonetischen Wortstruktur abzuleiten, die ich bereits vorher nachwies.

[3] Vgl. R. Jakobson, "Kindersprache, Aphasie und allgemeine Lautgesetze", (Uppsala, 1941); *Selected Writings*, I; *Phonological Studies* (The Hague, 1962), 328-401.

NORMALSPRACHIGE TRANSFORMATIONS-GESCHICHTEN UND KINDLICHE SATZBILDUNG

WILLEM KAPER

(Amsterdam)*

Unter NORMALSPRACHE soll in diesem Referat verstanden werden: die Sprache, die der "native speaker" als nicht abweichend von einer ihm unbewußten, aber intuitiv erkannten Norm empfindet. Der Begriff ist nicht fest umrissen, und neben einer Communis opinio über die Norm lassen sich soziale und individuelle Unterschiede feststellen.[1]

TRANSFORMATIONSGESCHICHTE ist ein Begriff der theoretischen Sprachwissenschaft, die sich mit einer Idealform der Normalsprache beschäftigt.[2] In der generativen Transformationsgrammatik hat bekanntlich jeder Satz eine Tiefenstruktur: eine hypothetische Struktur, in der die lexikalen Elemente mit der grammatischen Beziehung, in der sie zueinander stehen, die Bedeutung des Satzes unmittelbar zum Ausdruck bringen.[3] Ein dem System der Sprache entsprechender Satz, m.a.W. ein grammatischer Satz, weist in seiner wahrnehmbaren Struktur — der

* Amsterdam, Universität.

1 Der ursprünglich angegebene Titel "Normalsprachige Tiefenstrukturen und kindliche Satzbildung" erwies sich als weniger adäquat.

2 Vgl. Noam Chomsky, *Aspects of the Theory of Syntax*, (Cambridge, Mass., 1965), S. 3.

3 Noam Chomsky, *Deep Structure, Surface Structure, and Semantic Interpretation*. Reproduced by the Indiana University Linguistics Club, January, 1969, S. 3. Vgl. auch E. Vasiliu u. Sanda Golopenţia-Eretescu, *Sintaxa transformaţională a limbii române* (Bukarest, 1969), S. 24.

Oberflächenstruktur — gewöhnlich einen oder mehr Unterschiede gegenüber der Tiefenstruktur auf. Diese Unterschiede lassen sich theoretisch als Transformationen beschreiben, d.h. als Regeln, durch deren Anwendung eine Struktur in eine andere verwandelt wird.[4] Die Aufeinanderfolge dieser Transformationsregeln nennt Chomsky [5] "transformational history", Transformationsgeschichte.

Auf das von Chomsky [6] angeschnittene Problem, ob die Transformationen die Bedeutung eines Satzes beeinflussen können, kann jetzt nicht eingegangen werden. Wohl aber berührt sich das hier Vorzutragende mit der Frage, ob Tiefenstruktur und Transformationsgeschichte einer mentalen Realität entsprechen.[7] In der Satzbildung von Kindern im Vorschulalter gibt es nämlich Erscheinungen, die auf eine Beantwortung in positivem Sinn deuten könnten. Es ist klar, daß die Gültigkeit einer solchen Hypothese systematisch geprüft werden sollte. Dieses Referat soll dazu nur eine Anregung sein.

Betrachten wir einen deutschen Satz wie etwa:

(1) *Hans sagt, daß er zwei Butterbrote ißt.*

Ein solcher Satz ist in der Oberflächenstruktur doppeldeutig. Das kann dadurch explizit gemacht werden, daß man ihn als eine Konstruktion beschreibt, der zwei Tiefenstrukturen zugrunde liegen. Tiefenstrukturen sind *per definitionem* abstrakt, aber der Anschaulichkeit wegen werden sie hier teilweise konkretisiert:

(1a) *X sagt* + C
X ißt zwei Butterbrote
(1b) *X sagt* + C
Y ißt zwei Butterbrote.

[4] Noam Chomsky, *Aspects of the Theory of Syntax*, S. 16.-A. Kraak, *Negatieve zinnen: Een methodologische en grammatische analyse*, Diss. Amsterdam (Hilversum, 1966), S. 77f.
[5] Noam Chomsky, *Aspects of the Theory of Syntax*, S. 130.
[6] In Noam Chomsky, *Deep Structure, Surface Structure, and Semantic Interpretation.*
[7] Vgl. Noam Chomsky, *Language and Mind* (New York-Chicago-San Francisco-Atlanta, 1968), S. 16.

Durch Einbettung des jeweils zweiten Satzes in die durch C reservierte Stelle des ersten, und Einführung der Konjunktion mit der erforderlichen Wortumstellung entstehen die Strukturen:

(1c) *X sagt, daß X zwei Butterbrote ißt*
(1d) *X sagt, daß Y zwei Butterbrote ißt.*

Ersetzt man die Variablen X und Y durch Eigennamen, so ergibt das:

(1e) *Hans sagt, daß Hans zwei Butterbrote ißt*
(1f) *Hans sagt, daß Peter zwei Butterbrote ißt.*

Die doppeldeutige Oberflächenstruktur — wie (1) — wird dadurch generiert, daß man den jeweils zweiten Eigennamen durch ein Personalpronomen der dritten Person ersetzt. Die Pronominalisierungstransformation[8] ist obligatorisch, wenn der Eigenname im eingebetteten Satz mit dem des Matrixsatzes identisch ist.

Um die Verständigung zu erleichtern wurde ein deutsches Beispiel gewählt. Der entsprechende holländische Satz hat die gleiche Transformationsgeschichte, und deswegen ist es legitim, zum Vergleich Äußerungen von zwei holländischen Knaben heranzuziehen. Die Knaben sind Brüder, beide im Vorschulalter: Erik und der zwei Jahre und zwei Monate jüngere Hans.

Zunächst zwei Sätze vom Älteren der beiden:

(2) *Hans zegt, dat Hans er twéé eet* (E 5 ; 3,5)
(3) *Pappa, Hans zegt dat Hans het eerste klaar is* (E 5 ; 7,19). (Übersetzung S. 137 f.)

Die Konstruktion mit *er* in (2) ist grammatisch und interessiert in diesem Zusammenhang nicht.[9] Beachtenswert ist dagegen die Wiederholung des Eigennamens statt der Pronominalisierung im eingebetteten Satz. Wird hier doch in der Oberflächenstruktur sichtbar, was in der Beschreibung der entsprechenden normalsprachigen Sätze hypothetisch als Stufe der Transformations-

[8] Vgl. Manfred Bierwisch, "Grammatik des deutschen Verbs", *Studia Grammatica* II (Berlin, 1967), 96.
[9] Siehe jedoch unten, S. 135.

geschichte vorausgesetzt wird. Das berechtigt zu der Vermutung, daß das Kind auf der Stufe vor der Pronominalisierungstransformation stehengeblieben ist.

Eine verwandte Erscheinung findet sich bei Leopold[10] in bezug auf das Possessivpronomen. Seine Tochter Hildegard fragte mit 2½ Jahren:

(4) *What's the doggie got in the doggie's mouth?* (Hdg 2 ; 6,5)

Dieser Satz war eine Wiederholung in geänderter Form der unmittelbar vorher gestellten Frage:

(4a) *What's the doggie got in her mouth?*

Leopold schreibt die Veränderung der Formulierung der Tatsache zu, daß Hildegard das Inadäquate ihrer Äußerung fühlte. Die Form *his* war ihr aber noch nicht geläufig, da die Mutter weniger häufig von männlichen Lebewesen sprach. In diesem Licht ist die Wiederholung eines weiblichen Vornamens im nicht eingebetteten Satz interessant, die H um die Mitte seines fünften Lebensjahres hören ließ:

(5) *Pappa, weet je hoe laat Eesje jarig is? Eesje is in de zomer jarig* (H 4 ; 6,24).

Die beiden Knaben hatten nämlich keine Schwester und zeigten mehrmals, daß ihnen der Gebrauch des Personalpronomens der 3. Pers. Sg. weiblich in diesem Alter noch Schwierigkeiten bereitete.[11] Die Vermutung dürfte nicht unbegründet sein, daß zwischen Wiederholung des Eigennamens und Unsicherheit im Gebrauch des Pronomens irgendein Zusammenhang besteht.

Diese Hypothese wird gestützt durch ein Beispiel von H aus der Zeit, wo er zwar das Personalpronomen der 3. Pers.Sg. für männliche Personen schon richtig anwandte, die Objektsform

[10] Werner F. Leopold, *Speech Development of a Bilingual Child: A Linguist's Record IV* (Evanston, Ill., 1949), S. 28.

[11] Siehe W. Kaper, *Kindersprachforschung mit Hilfe des Kindes: Einige Erscheinungen der kindlichen Spracherwerbung erläutert im Lichte des vom Kinde gezeigten Interesses für Sprachliches* (Groningen, 1959), S. 152f.

aber noch nicht gleich fand. Nach einer Zögerung, die auf ein Suchen und Nichtfinden der richtigen Pronominalform deutet, setzte er nämlich den Namen *Pappa* ein. Er wollte mitteilen, daß der Vater ihm erlaubt hätte, zu ihm zu kommen, und sagte:

(6) *Mama, Papa zegt ik bij* ... (Zögerung) ... *ik bij Papa mag*! (H 2 ; 9,25).[12]

Das soll indessen nicht heißen, daß eine Wiederholung dieser Art nur dann eintritt, wenn das Kind die Pronominalform nicht finden kann. Die beiden ersten Beispiele (2) und (3), wo der Name *Hans* wiederholt wird, sprach E im 6. Jahr, wo die Form *hij* ihm längst geläufig war.[13] Es ist aber zu berücksichtigen, daß die Wiederholung hier im eingebetteten Satz erfolgte. In einer früheren Äußerung von E unterbleibt nämlich die Pronominalisierung im eingebetteten Satz, während sie im sich daran anschließenden Satz stattfindet. Diesmal betrifft es die 1. Pers.Sg.:

(7) *Mama denkt, dat Erik een huis maakt, maar ik maak een kerk* (E 4 ; 5,17).

Das Material reicht nicht aus, um zwingende Schlußfolgerungen zu ziehen. Als Arbeitshypothese ließe sich aber folgende Erklärung vorschlagen: das scheinbar inkonsequente Verfahren dürfte damit zusammenhängen, daß eine Einbettungstransformation mit der obligaten Wortumstellung komplizierter ist als bloße Koordination von Sätzen.[14]

Es gibt Beispiele anderer Art, wo gleichfalls die Tiefenstruktur der kindlichen Sätze derjenigen der Normalsprache entspricht, während die Transformationsgeschichte kürzer ist. Etwa drei Monate nach seinem fünften Geburtstag sagte H:

(8) *Net zo'n fles als deze fles* (H 5 ; 3,4).

[12] Siehe dazu W. Kaper, *Kindersprachforschung mit Hilfe des Kindes* (Groningen, 1959), S. 138f.

[13] Schon mit 3; 5, 24 wurde sie notiert.

[14] Die Frage, ob Koordination von Sätzen als 'Einbettung' beschrieben werden darf, ist in diesem Zusammenhang nur ein terminologisches Problem. Vgl. jedoch Simon C. Dik, *Coordination: Its Implications for the Theory of General Linguistics*, Diss. Amsterdam (Amsterdam, 1968), S. 97, Anm. 43.

Die Struktur dieses Satzes entspricht einer Stufe der Transformationsgeschichte, auf die in der Normalsprache eine Elisionstransformation folgen würde:

(8a) *Net zo'n fles als deze.*

Etwas verwickelter liegt die Sache bei einem von E mit 4½ Jahren gesprochenen Satz:

(9) *Welke nummer is deze nummer altijd?* (E 4 ; 6,15).

Das holländische Sprachsystem verlangt hier:

(9a) *Welk nummer is dit?*

Das logisch überflüssig erscheinende *altijd* ist in diesem Zusammenhang irrelevant, wie auch die Flexionsform *welke* statt *welk*. Worauf es ankommt, ist, daß im kindlichen Satz nicht nur das Substantiv nicht elidiert wird, sondern daß auch die Neutralisierung des Demonstrativs unterbleibt: die Ersetzung von *deze* durch *dit* ist hier im System der Normalsprache obligatorisch.[15]

Das Stehenbleiben auf der Stufe vor der Elisionstransformation läßt sich nicht nur beim Substantiv beobachten. H wiederholt zweimal ein zweites Partizip, wo Elision normal gewesen wäre:

(10) *Papa, 'k heb karnemelk gehad en chocola gehad en een beetje water gehad* (H 3 ; 8,1).

Zu einer anderen Erwägung führt folgende Äußerung, gleichfalls von H:

(11) *Weet je wat in me mond zit? Zànd in me mond!* (H 3 ; 8,5).

Hier hat schon Elision stattgefunden, denn in der Tiefenstruktur muß man eine VP annehmen. In der Normalsprache wäre eine weitere Reduktion — nämlich auf das bloße Substantiv — akzeptabel gewesen. Man kann sich aber auch eine andere Transformation denken, und zwar:

(11a) *zand zit in mijn mond* $\Rightarrow$ *er zit zand in.*

[15] H führte sie mit 4;1,19 durch in seiner Frage: *Mamma, welke pap is dit?*

Eine ähnliche Transformationsmöglichkeit bietet ein Satzpaar, das H mehr als ein Jahr früher (mit etwa 2½ Jahren) sprach:

(12) *Erik hebt een auto*; *ik hebt ook een auto* (H 2; 5, 20).

Auch hier ist eine *er*-Transformation erlaubt:

(12a) *ik heb ook een auto* ⇛ *ik heb er ook een.*

Es ist nicht der Zweck dieses Referats, auf den Unterschied zwischen den beiden *er*-Transformationen einzugehen;[16] was hier interessiert, ist, daß in der Normalsprache ein Element elidiert und *er* eingefügt werden kann.[17] Das Kind läßt diese Möglichkeit unbenutzt, was an sich nichts besagt. Es dürfte aber nicht uninteressant sein, zum Vergleich eine Konstruktion heranzuziehen, die H im Anfang seines sechsten Jahres bildete:

(13) *Erik hééft geen één paar*; *Erik heeft er zès paar*! (H 5 ; 1,12).

Dieses Beispiel zeigt einen weiteren Schritt: *er* ist eingefügt, aber das Substantiv ist nicht elidiert. Wieder liegt in der Oberflächenstruktur eine Transformationsstufe offen zutage, welche die Grammatik nur hypothetisch voraussetzt.

Die deutsche Umgangssprache kennt Konstruktionen, denen — nach der Hypothese der generativen Transformationsgrammatik — dieselbe Tiefenstruktur zugrunde liegt wie dem possessiven Genitiv, z.B.:

(14) *das Auto von Heinrich*
(14a) *dem Heinrich sein Auto.*

Bezeichnen wir ihre jeweilige Struktur — kurz aber ungenau — als 'Präpositionalkonstruktion' bzw. 'Dativkonstruktion'. Steht das regierende Substantiv mit dem unbestimmten Artikel, so ist nur die Präpositionalkonstruktion erlaubt, z.B.:

(14b) *ein Auto von Heinrich.*

16 Zum heiklen Problem der Funktionen von *er*, siehe Gunnar Bech, *Über das niederländische Adverbialpronomen ER* (= *TCLC* VIII) (Kopenhagen-Amsterdam, 1952).

17 Die Aussprache von *een* ändert sich bei der Transformation, was für das hier Ausgeführte irrelevant ist.

Die Dativkonstruktion ist ungrammatisch, wie z.B.:

(14c) **dem Heinrich ein Auto*
(14d) **ein dem Heinrich sein Auto.*

Entsprechende Strukturen mit der gleichen Kontextbeschränkung weist das holländische Sprachsystem auf. Diese Kontextbeschränkung läßt H unberücksichtigt, wenn er, gut vier Jahre alt, ausruft:

(15) *Daar is nòg een bus T zijn halte*! (H 4 ; 1,13).[18]

Zu vergleichen ist eine Äußerung, auch von H, die mehr als einen Monat später erfolgte:

(16) *'k Ga in jullie z'n kamer* (H 4 ; 2,25).

Hier gilt in der Normalsprache die Beschränkung, daß bei Pronominalisierung des abhängigen Substantivs nur die Präpositionalkonstruktion zulässig ist.[19] Auch die bekannte Erscheinung, daß Kinder oft den unbestimmten Artikel mit einem pluralischen Substantiv verbinden,[20] läßt sich als Außerachtlassen einer Kontextbeschränkung beschreiben.

Fassen wir zusammen. Wir haben festgestellt, daß bisweilen Stufen der Transformationsgeschichte, die nach der Hypothese der generativen Transformationsgrammatik den Oberflächenstrukturen der Normalsprache zugrunde liegen, in der Kindersprache als Oberflächenstruktur erscheinen.[21] Diese Beobachtung könnte

[18] Eine mögliche Erklärung ist, daß *bus T zijn halte* vom Kind als ein Ganzes gehandhabt wird, wie auch *één kind zijn tand* in *Pappa, één kind zijn tand van mijn school is eruit gevallen* (H 4;4,4). Siehe zu diesem Problem: W. Kaper, "Kindersprache und 'linguistic competence'", in *Lingua* 21 (1968), 230f.

[19] Vgl. auch *Hier is nou de beesten zijn bed* (H 5;7,27), wo das Possessivpronomen die Singularform hat, obgleich im Kontext ein pluralisches Substantiv steht.

[20] Z.B. *Butje heb een eigen jampotten* (E 2;10,29).

[21] Als äußerste Konsequenz könnte man sich denken, daß in den Anfängen des kindlichen Sprechens die Oberflächenstruktur unmittelbar die Tiefenstruktur repräsentiert. Vgl. David McNeill, "Developmental Psycholinguistics", *The Genesis of Language: A Psycholinguistic Approach*, ed. by Frank Smith and George A. Miller (Cambridge, Mass., and London, England, 1966), 51: "it is not too unreasonable to think of children 'talking' base strings directly". Zum Begriff "base string" ("basic string"), siehe Noam Chomsky, *Aspects of the Theory of Syntax*, S. 17.

beitragen zur positiven Beantwortung der Frage, ob Tiefenstruktur und Transformationen tatsächlich zum kognitiven System gehören, das Chomsky[22] als "linguistic competence" beschreibt. Überschauen wir die Beispielsätze, so zeigt sich, daß die Stufe, die sie jeweils darstellen, einfacher und weniger differenziert ist als die der entsprechenden normalsprachigen Sätze: es fehlt die Pronominalisierung, die zur Unterscheidung nach Person, Geschlecht und Kasus[23] zwingt; es fehlt die Elision und damit die Entscheidung, was elidiert und ob neutralisiert werden soll; es fehlt die *er*-Transformation mit ihren verzwickten Möglichkeiten; oder es werden Kontextbeschränkungen außer acht gelassen, die als solche differenzierend gewirkt hätten. Ob die Beispiele repräsentativ sind, wäre noch zu erforschen. Auf jeden Fall stehen sie im Einklang mit Jakobsons[24] Entdeckung, daß das Sprachsystem dem Grundsatz des maximalen Kontrastes gehorcht, und vom Einfachen und Ungegliederten zum Abgestuften und Differenzierten vorschreitet.

Übersetzung der holländischen Sätze

(2) Hans sagt, daß Hans (deren) zwei ißt.
NB: Im Holländischen ist *er* ('*deren*') obligat.
(3) Papa, Hans sagt, daß Hans zuerst fertig ist.
(5) Papa, weißt du, wie spät [d.h. um wieviel Uhr] Eesje Geburtstag hat? Eesje hat im Sommer Geburtstag.
(6) Mama, Papa sagt, ich zu ... [Zögerung] ... ich zu Papa darf!
(7) Mama denkt [d.h. glaubt], daß Erik ein Haus macht, aber ich mache eine Kirche.
(8) Genau so'ne Flasche wie diese Flasche.
(8a) Genau so'ne Flasche wie diese.
(9) Welche Nummer ist diese Nummer immer?
(9a) Welche Nummer ist dies?
(10) Papa, ich habe Buttermilch gehabt und Schokolade gehabt und ein bißchen Wasser gehabt.

[22] Noam Chomsky, *Language and Mind*, S. 4.
[23] Für die holländische Sprache kommt nur Subjekts- und Objektskasus in Betracht.
[24] Roman Jakobson, *Kindersprache, Aphasie und allgemeine Lautgesetze* (Uppsala, 1941), S. 54.

(11) Weißt du, was in meinem Mund sitzt [d.h. ist]? Sànd in meinem Mund.

(11[a]) Sand sitzt in meinem Mund ⇒ es sitzt Sand darin.

(12) Erik habt ein Auto; ich habt auch ein Auto.

(12[a]) Ich habe auch ein Auto ⇒ ich habe (deren) auch eins.
NB: Im Holländischen ist *er* ('deren') obligat.

(13) Erik hàt nicht éin Paar; Erik hat (deren) sèchs Paar!
NB: Wenn *er* ('deren') gebraucht wird, soll im Holländischen *paar* elidiert werden.

(15) Da ist nòch eine (dem) Bus T seine Haltestelle!

(16) Ich gehe in (dem) euch sein Zimmer.

Anm. 15. Mama, welcher Brei ist dies?

Anm. 18. Papa, einem Kind sein Zahn von meiner Schule ist herausgefallen.

Anm. 19. Hier ist nun den Tieren sein Bett.

Anm. 20. Butje hab eine eigene Marmeladengläser.
NB: Mit *Butje* bezeichnete Erik sich selbst.

DISCUSSION:

MIKEŠ: Die Entwicklung der Syntax in der Kindersprache geht in der Richtung von Oberflächenstruktur identisch mit der Tiefenstruktur, zur Oberflächenstruktur nicht identisch mit der Tiefenstruktur. Im Vorschulalter kann die Kompetenz der Kinder geprüft werden, um diese Hypothese zu beweisen. Ein gutes Beispiel dafür ist Carol Chomský's Buch.

OKSAAR: Methodisch ist es nicht richtig, die gesprochene Sprache der Kinder auf der Basis der geschriebenen Sprache der Erwachsenen zu analysieren (die normative Grammatik baut ja auf dieser auf). Wir müssen die gesprochene Sprache der Umgebung des Kindes als Vergleich heranziehen. Man darf auch nicht vergessen, daß die Tiefenstruktur hypothetischen Charakter hat und diese hypothetischen Tiefenstrukturen müssen sehr genau überprüft werden, ehe man auf ihnen aufbaut.

RŪĶE-DRAVIŅA: Für die Feststellung des speziell Kindersprachlichen sollte man die Sprache der Kinder mit der gesprochenen Sprache (und nicht mit der normierten Sprache) vergleichen, vor allem aber mit der Sprache der näheren Umgebung des Kindes. Die Pronominalisierung ist eine schwierige Aufgabe für das kleine Kind: die pronominale Transformation betrifft auch die persönlichen und possessiven Pronomina (Eigenname anstatt des Pronomens). Verwechslungen von *du* und *ich* u.a. Es gibt Parallelen aus anderen Sprachen (z.B. Lett.), die die Hypothese von W. Kaper stützen.

LA CONSTATATION D'OPPOSITIONS PHONÉMIQUES DANS LE SYSTÈME LINGUISTIQUE DE L'ENFANT

MARTIN KLOSTER-JENSEN

(Bergen)

L'aspect le plus important dans le langage enfantin, c'est celui du développement. Nous savons, à peu près, où aboutiront les efforts de l'enfant apprenant la langue maternelle, mais nous sommes bien moins renseignés sur les étapes de ce progrès. Il est un des mérites du président de notre colloque d'avoir fait avancer l'étude sur l'évolution linguistique de l'homme, surtout sur la maîtrise des symboles de la langue parlée.

L'enfant se voit plongé dans un milieu parlant. Avec ses semblables, il cherche à se servir des moyens de communication que lui offrent les entourages. Étant donné que l'enfant s'assimile au milieu linguistique sans autre introduction que les contacts naturels et de circonstance, nous devons supposer que la maîtrise de la langue maternelle repose uniquement sur l'analyse que fait chaque sujet parlant d'après les expériences qu'il a faites lui-même avec les moyens de communication à sa disposition. C'est pourquoi, pour découvrir le procès de l'acquisition de la langue maternelle, nous obseivons le parler d'un enfant en le comparant tout le temps avec la langue modèle des adultes. Toute différence entre les deux systèmes d'expression est considérée et mesurée comme une imperfection du langage enfantin. A notre avis, cette façon de raisonner est justifiée par le besoin de normes. Aussi est-ce aux normes que nous rapportons toute comparaison pour suivre le procès d'apprentissage de la langue maternelle.

Les normes sont de plusieurs catégories; elles se rapportent à différents niveaux de la hiérarchie des unités linguistiques (phonèmes, morphèmes, mots, phrases). Les normes portent également sur le SENS, c.-à-d. sur l'EMPLOI des symboles (signes); elles concernent tout comportement verbal.

Jetons un coup d'œil sur l'acquisition des unités 'vides' de la langue parlée, les phonèmes. S'il est vrai que l'enfant apprend d'abord à produire des complexes acoustiques qui correspondent aux mots et aux phrases, il n'en est pas moins sûr qu'il se produit un apprentissage des phonèmes, au fur et à mesure que le sujet parlant maîtrise la substitution génératrice.

On sait que l'enfant distingue les mots ENTENDUS bien longtemps avant d'être à même de les prononcer d'une façon différenciée. Cette observation trouve un parallèle dans l'*Informatorium mateřské školy* [de l'école-mère] de Comenius, chap. IV, 12,2, où la compréhension et la production de la parole sont présentées comme deux activités différentes. Cette division en deux de la maîtrise d'un idiome amène la conclusion qu'il sera utile de considérer deux systèmes phonémiques dans le langage enfantin. Les deux systèmes relèvent de la capacité de l'enfant à différencier, dans les paires de mots à opposition phonémique minima:

(1) en écoutant parler un adulte
(2) en s'exprimant par la parole.

Cette manière de présenter rappelle l'aspect double des vocabulaires: passif et actif, bien que la correspondance ne soit pas complète. Tandis qu'on a pu constater un accord raisonnable et, croyons-nous, généralement valable, pour la relation entre les vocabulaires passif et actif (grosso modo 3 à 1), nous savons encore très peu sur la maîtrise (passive ou active) des systèmes phonémiques. Il y a 18 ans que nous avons proposé de donner un nom spécial aux unités du système phonémique actif,[1] en présupposant que ce système actif garde son indépendance et continue à fonctionner même après une correspondance complète obtenue avec le système maîtrisé de façon passive. L'existence indépen-

[1] "Un pendant à la phonématique", *Universitetet i Bergen, Aarbok*: *Historisk-antikvarisk rekke* 4 (1952), 4.

dante de l'un et de l'autre système est prouvée par les cas d'aphasie d'ordre sensoriel ou moteur, et par le langage des sourds-muets. Nous avons proposé d'appeler les unités distinctives du système actif (moteur): KINÈMES.

Pour suivre maintenant le procès de l'apprentissage de la langue au niveau des phonèmes, nous aurons une tâche double:

(1) trouver les unités distinctives minima maîtrisées de façon 'passive'
(2) établir le système des KINÈMES, c.-à-d. des unités maîtrisées activement, et de façon motrice.

La différence de structure entre ces deux systèmes doit nous intéresser autant que celle entre les vocabulaires actif et passif.[2]

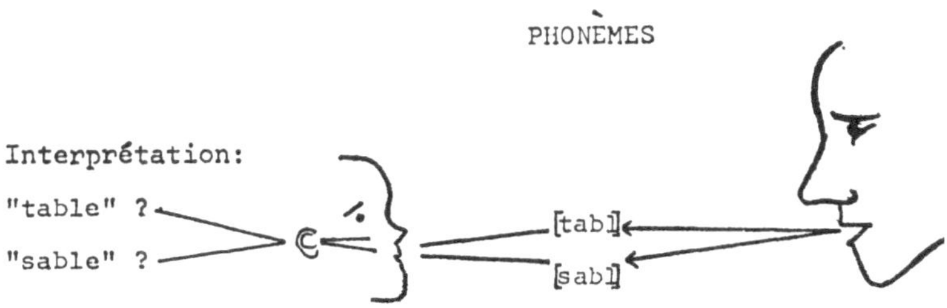

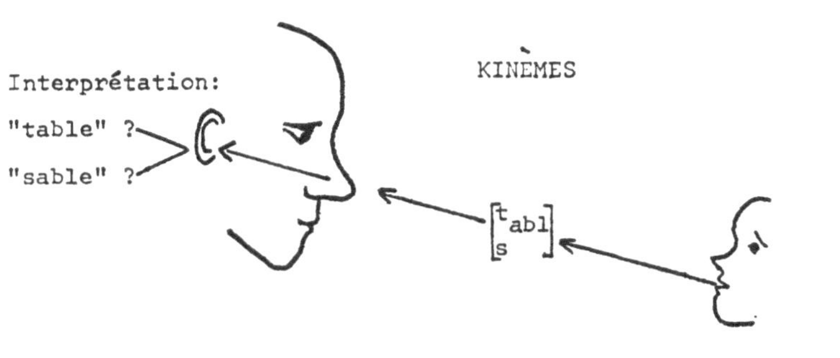

[2] L'intervention de Madame Mikeš a souligné encore que le système 'kinématique' est compliqué par les limitations distributionnelles qui s'imposent le plus souvent au début de l'apprentissage.

La technique pour établir les deux systèmes au niveau des unités vides[3] est différente selon qu'il s'agit de phonèmes ('passifs') ou de kinèmes. Dans ce dernier cas, la décision de pareil ou différent, fondamentale dans l'analyse de toute expression, doit être prise par le chercheur. Dans le cas de phonèmes (au sens spécialisé d'unités à interprétation auditive) c'est l'enfant qui décide. Prenons la paire minima TABLE ~ SABLE. S'il y a dans le système phonémique de l'enfant deux unités [t] et [s], il comprend les deux mots. Sinon, il les confond. De même, s'il y a dans son système deux kinèmes [t] et [s], il peut exprimer par les symboles de la langue parlée les deux concepts 'table' et 'sable'. Autrement, il les confond en parlant.

Ce n'est qu'après avoir établi une technique appropriée de recherche portant sur les questions mentionnées, qu'il sera possible d'attaquer le problème d'un ordre de rang des oppositions linguistiques. Cette remarque s'applique également à la fonction des traits distinctifs des unités du niveau phonémique.[4]

CONCLUSION: Le processus d'établir l'inventaire des kinèmes et des phonèmes dans le langage d'un enfant ne peut être accompli qu'entre l'enfant et le chercheur. Les rôles de "speaker" et d'interprétateur seront changés selon le système à déterminer.

DISCUSSION:

BOREL-MAISONNY: Cette distinction en kinèmes et phonèmes est applicable aux enfants ayant du retard de parole et introduirait une simplification. Il faudrait néanmoins tenir compte des troubles perceptifs qui peuvent modifier les rapports de ces facteurs.

MIKEŠ: A côté d'un système phonémique corrélatif il existe un système associatif: la théorie de Martinet appliquée au langage enfantin.

PAVLOVIĆ: Les observations de M. Kloster-Jensen touchent au problème lin-

[3] Louis Hjelmslev a créé pour toute unité linguistique à fonction purement différenciatrice, et vide de sens (phonème, graphème et autre) le terme *cénème*, de grec kenós 'vide'.

[4] Voir à ce sujet Sandra Hutcheson "Some Quantitative and Qualitative Criteria in Articulation Test Scoring", *The British Journal of Disorders of Communication* 3:1 (1968).

guistique le plus important, celui d'opposition phonologique. Non seulement les faits, mais aussi la manière de suivre les processus de stabilisation et d'interprétation minutieuse se présentent comme modèle pour les études des systèmes complexes.

SÖDERBERGH: It seems evident to me that prof. Kloster-Jensen's theory that the child knows the adult phonematic system before he is able to express it must be right. One example from my studies confirms this. A child mixes up the Swedish phonemes /ç/ and /s/ pronouncing the Swedish words /çɷ:l/ 'frock' and /sɷ:l/ 'sun' in the same way. When she one day succeeds in saying /çɷ:l/ correctly, she becomes very happy saying: "Oh, now I can say /çɷ:l/", and then she cries out a great many words all containing /ç/, which shows that she has been STORING these words in her memory, knowing that she has not been able to pronounce them correctly.

EINFLUSS EINIGER WICHTIGEN GESELLSCHAFTSFAKTOREN AUF DIE ENTWICKLUNG DER RICHTIGEN ARTIKULATION DER KINDER

VLADISLAV LEJSKA

(Kyjov)

In Jahren 1965-1966 wurde von uns die Nachforschung des Aıtikulationszustandes bei Kindern der ersten Klassen im Kreis Hodonín durchgeführt. Das versammelte Material dienste uns als Grundlage zur Auswertung von aktuellsten ganzgesellschaftlichen etiologischen Faktoren des festgestellten Stammelns.

Mit dem Einfluss von sozialen Verhältnissen auf die Entwicklung der Sprache und der Aussprache bei Kindern beschäftigte sich bei uns Štulíková*. Ihre Arbeit, die im Jahre 1934 publiziert wurde, bleibt bisher die einzige in tschechischer Fachliteratur, deren Aufgabe ähnliche Problematik zu lösen war. Es muss jedoch bemerkt werden, dass die damaligen sozialen Verhältnisse vom heutigen Lebensniveau unserer Familien bedeutend unterschiedlich waren. Seit der Zeit veränderte sich auch die Qualität und der Umfang der gesamten Gesundheitspflege. Die Hauptbetonung des Gesundheitsinteresses wird heutzutage, praktisch in allen medizinischen Gebieten, auf die Prevention gelegt werden. Zu den Fächern, deren Haupttätigkeit die Verhütung von verschiedensten Anomalien und Störungen bildet, gehört auch die Phoniatrie. Im Laufe der letzten Jahren gelang diese speziale Pflege bei uns zu einem grossen Aufschwung. Diese Tatsache gehört auch zu den positiven sozialen Einflüssen und Veränderungen und das nicht nur in ihrer präventiven sondern auch therapeutischen Tätigkeit. Die Ergebnisse von Štulíková bleiben für uns immerhin sehr interessant und belehrend und wir wollen sie deshalb, wenigs-

* Ztschr. *Úchylná mládež*, Praha 1934, 60-64

tens in einigen Punkten, mit unseren Schlussfolgerungen vergleichen.

Wir haben insgesamt 844 Kinder in 33 Schulen untersucht. Es handelte sich um Kinder sowohl aus kleinen Dorfschulen, mit drei Klassen maximal, wobei in einem Lehrraum abteilungsweise unterrichtet werden muss, so wie um Schulkinder aus kompletten neunklassigen Grundschulen in grösseren Gebietsortschaften und auch Kreisstädten. Die Untersuchung wurde zweimal durchgeführt und zwar zum Schulbeginn und am Ende des Schuljahres. Das Stammeln wurde bei 36% von Kindern festgestellt. Die Abweichungen an Frequenz in einzelnen Ortschaften waren in statistischer Hinsicht unbedeutend. Am häufigsten kam das Stammeln bei Zischlauten vor, und zwar in 42%, beim Schwingzischlaut *Ř* (z. B. *moře*) in 40%. In den meisten, dieses Problem schon früher behandelnden statistischen Arbeiten (Seemann, Ohnesorg und andere), wird an erster Stelle das Stammeln des Schwingzischlauts *Ř* angeführt und die Artikulationsstörungen der Zischlaute erstreckt an der zweiten Stelle oder noch später. In dieser Hinsicht kamen wir zu anderen Ergebnissen. Wir halten es für nichts Merkwürdiges, weil die Frequenz einzelner Stammellaute von vielen Faktoren abhängig ist, also auch von den Befunden an Sprachorganen (Dentition, Adenoides veget., usw.) das heisst also von der Qualität der damit zusammenhängenden ärztlichen Pflege. Die Ergebnisse können daher an einzelnen Schulen, in einzelnen Gebieten und Jahrgängen voneinander unterschiedlich sein.

Aus der reichen und bunten Palette der Gesellschaftsfaktoren, die den entscheidenden Einfluss auf die Entwicklung der richtigen Artikulation der Kinder ausüben können, haben wir drei Grundprobleme herausentnommen:

(1) Einfluss des Kindergartens,
(2) Einfluss der Volksschule,
(3) der Einfluss des Familienmilieus mit besonderer Rücksicht auf die Schulbildungsgrad der Eltern.

Alle 844 Schüler wurden von uns in drei Gruppen verteilt, je nachdem, welche Schule sie besuchen: entweder eine kleine

Volksschule mit geringer, das heisst unvollständiger Klassenzahl, oder eine komplette neunklassige Grundschule, oderaber eine moderne Schule in einer Stadt. An allen Gruppen haben wir die Frequenz des Stammelns nachgeforscht und festgesetzt. Auf Grund dieser Arbeit wurde festgestellt, dass zu Beginn des Schuljahres das maximale Vorkommen von Stammeln am häufigsten bei Schülern der undifferenzierten Klassen, das heisst in kleinen Dörfern — es war 45,7 %. Diese Kinder besuchten keinen Kindergarten. In solchen kleinen Dörfern gibt es nur die Möglichkeit in Sommermonaten einen wirtschaftlichen Hort zu besuchen, in dem die Kinder von unqualifizierten Arbeitskräften aus den Reihen der Eingeborenen betreut werden.

In grösseren Gemeinden mit einer kompletten neunklassigen Grundschule, besuchte keinen Kindergarten nur der kleinere Teil von Kindern und das Auftreten des Stammelns ist daher auch kleiner — nur 41 %. In den Städten besuchten fast alle Kinder den Kindergarten und das Vorkommen von Stammeln ist hier am geringsten — nur 28 %.

Der Einfluss des Kindergartens auf die Artikulation des Kindes ist also positiv und der Kindergartenbesuch erweist sich daher als sehr notwendig. Dies wurde bereits auch von Filcíková-Herfortová bestätigt und ihr Nachschlagebuch für Kindergärtenlehrerinnen bildet eine gute Hilfe bei der Einflussnahme auf die richtige Artikulation bei Kindern im vorschulpflichtigen Alter. Die Geltendmachung der, wenigstens elementaren, logopaedischen Standpunkte im Erziehungsprozess seitens der Lehrerinnen und Erzieherinnen, bildet die Bedingung deren erfolgreichen Arbeit.

Ferner wollten wir feststellen was für einen Einfluss auf die Entwicklung korrekter-Schriftartikulation bei Kindern die Unwelt der neunklassigen Grundschule ausübt. Bei der Lösung dieses Problems wurde folgendermassen vorgegangen:
zuerst haben wir in allen Schulen zu Beginn des ersten Jahrganges die Frequenz einzelner Stammellaute bei Kindern festgestellt. Im Laufe des Schuljahres besuchten die Kinder keinen logopaedischen Fachmann, sodass die Regelung der Artikulation spontan durchlief. Durch Analyse des Artikulationszustandes am Ende des

Schuljahres, haben wir bei denselben Kindern, die folgende Schlüsse gezogen (Abb. 1).

STAMMELN	SCHULE	VERBESSERUNG %
SIGMATIS.	komplett differ.	33
	wenigklass.	60
Ř + R	alle Schulen	24
Ď, Ť, Ň	komplett differ.	50
	wenigklass.	75

In kleinen Schulen mit unvollständiger Klassenzahl tritt das Stammeln schneller zurück, als in den grösseren vollkommen differenzierten Schulen. Man muss hier die Beeinflussung der richtigen Artikulation durch die älteren Mitschüler, die in demselben Klassenzimmer arbeiten, zugeben. (In den allermeisten Fällen arbeiten gemeinsam der erste und der dritte Jahrgang.) In einem Klassenzimmer gibt es, zusammengezählt, meistens etwa 30 Schulkinder. Der positive Einfluss älterer Mitschüler in wenigklassigen Schulen war besonders bei der Artikulation von Zischlauten und weichen Lauten *Ď*, *Ť*, *Ň* bemerkbar. Auf die Regelung richtiger Artikulation von Schwingzischlauten war weniger markant. Der spontane Rückgang von Rhotazismen durchläuft analog wie in den höher organisierten Schulen.

Von weiteren gesellschaftlichen Faktoren, die noch heute auf die Entwicklung der Artikulation bei Kindern Einfluss ausüben könnten, beachteten wir das Familienmilieu der Schüler unter besonderer Berücksichtigung des Bildungsgrades ihrer Eltern. Je nach dem Bildungsgrade der Eltern, teilen wir die soziale Familienlage unserer Schüler in drei Gruppen:

(1) Beide, Vater und Mutter, haben eine Grundschulbildung;
(2) Wenigstens einer von den Eltern hat eine Mittelschulbildung;
(3) Wenigstens einer von den Eltern hat eine Hochschulbildung.

Die ganze Zusammenstellung der 844 verfolgten Kinder verteilten wir in zwei Gruppen und das — auf Kinder mit dem Stammeln (insgesamt 304) und auf richtig sprechende Kinder (insgesamt 540). Ferner werden absichtlich die Dorfschulen (die wenigklassigen und auch die neunklassigen Grundschulen), von den städtischen Schulen abgesondert und man zieht folgende Schlüsse:

In 18 Dorfschulen gab es zu Beginn des Schuljahres in den ersten Klassen insgesamt 470 Schüler und davon wurde bei 199 Kindern die Störung der Artikulation festgestellt. Die übrigen 271 Schüler artikulierten schon alle Laute richtig.

Die Eltern der 271 richtig sprechenden Schulkindern sind, ihrem Bildungsgrad nach, folgendermassen in drei Gruppen verteilt:

Eltern	Anzahl	%
1. Grundschulbildung	185	68,27
2. Mittelschulbildung	64	
3. Hochschulbildung	22	31,73

Man könnte daher das Verhältnis zwischen einzelnen Gruppen gekürzt mit 8,5 : 3 : 1 zum Ausdruck bringen.

Die Eltern der 199 Kinder mit Stammeln haben wir, wieder je nach ihrem Bildungsgrade, wie folgt verteilt:

Eltern	Anzahl	%
1. Grundschulbildung	178	89,45
2. Mittelschulbildung	21	10,55
3. Hochschulbildung	0	0

Das gegenseitige Verhältnis kann man wieder gekürzt mit 8 : 1 : 0 ausdrücken.

In der Gruppe der Familien mit stammelnden Kindern ist das Niveau des Bildungsgrades niedriger als bei Eltern mit richtig sprechenden Kindern. Es kann also gesagt werden, das die Bildung, der auf dem Lande lebenden Eltern, sich als ein hochbedeutender Faktor mit Bezug auf das Stammeln bei Kindern erweist: $X^2 = 29{,}28$; $P < 0{,}01$.

In den Stadtschulen war der Bildungsgrad bei Eltern — beider Kindergruppen — der stammelnden und der richtig artikulierdenden — praktisch gleich hoch und kann im Verhältnis von 6 : 3 : 1 ausgedrückt werden.

In den Städten wurden insgesamt 374 Schulkinder untersucht. Davon gab es 105 Kinder mit Artikulationstörung. Die Bildungsgrad bei Eltern der 269 richtig sprechenden Kinder war wie folgt:

Eltern	Anzahl — Famil.	%
1. Grundschulbildung	162	60,22
2. Mittelschulbildung	80	29,74
3. Hochschulbildung	27	10,04

Die Bildung bei Eltern der 105 stammelnden Schulkindern:

Eltern	Anzahl — Famil.	%
1. Grundschulbildung	63	60,00
2. Mittelschulbildung	32	30,48
3. Hochschulbildung	10	9,54

Bei der städtischen Population erweist sich die Bildung der Eltern in Bezug auf das Stammeln bei Kindern als kein Faktor von Belang: $X^2 = 0{,}03$; $P > 0{,}05$.

ZUSAMMENFASSUNG

(1) Der Einfluss des Kindergartens auf die Artikulation des Kindes ist positiv und der Kindergartenbesuch erweist sich daher als sehr notwendig. Die Geltendmachung der, wenigstens elementaren logopaedischen Standpunkte im Erziehungsprozess seitens der Lehrerinnen und Erzieherinnen, bildet die Bedingung deren erfolgreichen Arbeit.

(2) In kleinen Schulen mit unvollständiger Klassenzahl tritt das Stammeln schneller zurück, als in den grösseren vollkommen differenzierten Schulen. Man muss hier die Beeinflussung der richtigen Artikulation durch die älteren Mitschüler, die in demselben Klassenzimmer arbeiten, zugeben.

(3) In den Städten entscheidet nicht die Schulbildung der Eltern in der Entwicklung der richtigen Artikulation der Kinder. Es ist dadurch erklärlich, dass auch unsere Arbeiterfamilien in den meisten Fällen in Städten in gesellschaftlicher Höhe sind, was einen positiven Einfluss auf die Entwicklung der Sprache hat.

(4) Auf dem Lande überleben noch immer der Bildungsfaktor und dadurch auch die soziale Lage der Eltern als eine auf die Sprachentwicklung des Kindes Einfluss wirkende Kraft. Wir nehmen aber nicht an, dass das grössere Auftreten von Stammeln bei Kindern nur durch die niedrigere Bildung der Eltern verursacht wird, sondern auch durch andere Faktoren charakteristisch für Leben auf dem Lande.

SCHLUSSFOLGERUNG

Der Einfluss der Gesellschaftsfaktoren auf die Artikulation der Kindersprache ändert sich, so wie im gesellschaftlichen Leben selbst und in seinem sozialen Niveau Veränderungen vorgehen. Bei einigen Einflüssen ist eine Veränderung zu Gunsten des von uns gelösenen Problems eingetreten (hohes soziales Niveau unserer Familien), bei anderen zu Ungunsten (hoher Beschäftigungsgrad bei Müttern), auch dann wenn es sich zeigt, dass ein guter Erziehungsprozess im Kindergarten diese typisch tschechoslowakische Erscheinung nachholen und kompensieren kann.

Wir sind uns dessen bewusst, dass einige Schlüsse in dieser Arbeit gewiss nur für unsere Zusammenstellung typisch bleiben. Die anderen jedoch werden auch zur allgemeinen Geltung gebracht und das sowohl im rein informativen als auch praktischen Sinn.

DISCUSSION:

DVONČOVÁ: Ich habe die Frage, ob man feststellen kann, welcher von diesen drei im Referat genannten Grundeinflüssen sich als der wichtigste für die Entwicklung der Kindersprache zeigt.

ANTWORT: Es ist schwer zu sagen, welcher von den drei genannten Gesellschaftsfaktoren der wichtigste ist. Mit diesem Problem haben wir uns in unserer Arbeit nicht beschäftigt. Ich bin überzeugt, dass alle gleich wichtig sind.

RŪĶE-DRAVIŅA: Im Vortrag wurde von dem 'Stammeln' gesprochen, obwohl man eigentlich — allem Anschein nach — auch die Abweichungen von den Ausprachenormen der Standardsprache meinte. Im Grunde sind es zwei ganz verschiedene Phänomene, die nicht von denselben Faktoren abhängen.

SIERTSEMA: There may be other causes for stuttering. In Africa a research among primary-school children proved that it was especially frequent among children from families built up according to new western pattern, in which neither the husband wanted to take responsibility of the child (as he ought to do by the new pattern rules) nor the wife's brother (old pattern), so the child felt insecure.

ARTICULATORY DISORDERS AND THEIR EFFECT IN COMMUNITY

JOZEF LIŠKA

(Košice)

My paper draws attention to the problems of articulatory disorders, why and how they affect the child's personality and what is their effect in community.

There are many lisping children and many of those who mispronounce the *R* sound. Fortunately most of the children with mild articulation disorders do not even know about it, or do not worry about them.

There are some individuals, however, whose personality might be affected even by a slight articulatory defect. This may be associated with such factors as the emotional life and maladjustment of the individual on one hand, or the environment and the way in which such an individual is treated in the community on the other hand.[1] But it has been found that severe articulation disorders lead to personality deviations and internal emotional disturbances with every child.[2] Some of these children are shy, timid, lacking in achievement and with anxiety which leads to isolation, while the others are unsteady, restless, illtempered, tiresome, negativistic and aggressive.

I want to discuss these problems from the point of view of phonetics, phonology, the speech communication process and psychology.

[1] Margaret Hall Powers, "Clinical and Educational Procedures in Functional Disorders of Articulation", in L. E. Travis, *Handbook of Speech Pathology* (New York, 1957), 770.

[2] Jozef Liška, *Poruchy reči* (Bratislava, 1957), pp. 28-29.

Human speech is a very complicated process. To master this tool it means a laborious and long term operation.[3] This process is more complicated in children with language disturbance, in our case, with articulation defects.

Very often a child's speech learning is influenced by his environment and especially by his parents' faulty, improper speech patterns. Parents who imitate misarticulation after their children contribute thus to fixing incorrect acoustic and motor articulation patterns.

In normal language learning in child since his babbling stage a correlation between speech sounds and the taction-kinesthesis (and their feedback) is being established.[4] Both these activities are closely related to the visual activity. We assume that both the phonemic (phonological) system and the kinesthetic one function as equal in child's language.[5,6] They form a unity — we may call it 'suprasystem' — when completed with the visual system. There must be a perfect automatic coordination of these systems if the language is to work normally.

I would suggest addental lisping (addental sigmatism) as the simplest example of the articulatory disorder in the Slovak (Czech, German or French) language. Let us refer to the Standard Slovak system by the letter A and the system with the defect articulation by the letter B.

What is the difference between the system A and B?

Addental pronunciaton of the Slovak [s] means that there has

[3] Karel Ohnesorg, *O mluvním vývoji dítěte* (Praha, 1947), pp. 9-10; *Idem, Fonetická studie o dětské řeči* (Praha, 1948), pp. 64-68; *Idem, Druhá fonetická studie o dětské řeči* (Praha, 1959), p. 76.

[4] Jozef Liška, "Kodifikácia spisovnej výslovnosti", *Kultúra spisovnej slovenčiny* (Bratislava, 1967), 157.

[5] A. M. Liberman, "Some Results of Research on Speech Perception", *J.Acoust.Soc.Amer.* 29 (1957), 117-123. He thinks that perception of speech sounds is somehow more closely related to the articulation than to the acoustic stimulus. He gives priority to the motokinesthetic feedback loops in the comprehension-use of language.

[6] R. Brain says that since motor patterns are the earliest events in the comprehension of speech, neurological processes of kinesthetic feedback must mediate between the acoustic stimulus and its perception. *Speech Disorders* (Washington, 1961), pp. 80-83.

been a phonetic change both articulatory and acoustic. It is because of both articulatory and acoustic qualities that the addental [s] does not correspond to the Standard Slovak pronunciation. It is a disturbing but not a confusing sound here. In other words, from the phonetic point of view there is a difference between A and B systems.

With the articulation disorders of a phonetic character the amount of disturbance in the communication is specified by the size of the disrupting signal and the level of noise in the transmission channel. For instance the volume of noise with the nasal sigmatism (lisp) is substantially greater than that with the addental or interdental sigmatism. This expresses not only the noise level of the transmitting message but also the measure of its intelligibility, as well as the size of the psychical tension between the source B and the receiver A.

The articulation disorders with the phonetic character do not have such bad effects as those which cause the phonemic (phonological) change of the standard system A.

The phonological system of the Standard Slovak is not affected by the change of the Standard Slovak [s] into addental [s]. It means that there is no difference between B a A system from the phonological (phonemic) point of view. Addental [s] is meant as an articulatory variant of the sound, which does not bring the danger of confusion with any other phoneme in Standard Slovak, nor does it bring the distinction between words and morphemes.

But this might not be the case in some other languages. For instance, in English the substitution of the dental-alveolar [s] sound by the addental [s] means also change in the phonological (phonemic) system of the language bringing confusion with the phoneme [θ] (= *th* in orthography). In this case there is no distinction between the words *sick* — *thick* [sik — θik], *pass* — *path* [pa:s — pa:θ].

A very simple example of the articulation disorder that illustrates the change of the phonological system in Slovak (as well as in English) is the identical pronunciation of [l] and [r] phonemes. It means that there is only one motor pattern representing two phonemes in the motor area of the CNS. It does not make any distinction between words *lak/rak*, *Milo/Miro*, *dal/dar* in Slovak; *low/row*, *lead/read* in English.

In the sensory area of the CNS two situations at least may result:

(1) In the first case one acoustic pattern is established for both [l] and [r] phoneme. The child perceives not only the production of his own identical [l/r] sound as one sound, but also two different [l] and [r] sounds coming from his environment. At school the child confuses *l* and *r* letters in his writing in this case.

(2) In the second case two different acoustic patterns are established, e.g. one for [l] and the other for [r] phoneme which correlates with one motor pattern [l/r] in his motor area of the CNS. Sometimes, of course, the matter is more complicated as the child in all language learning is dependent upon correlated power of auditory, visual and tactile-kinesthetic feedback.

From the point of view of the speech communication process during the transmission of a distorted speech signal, that means during coding and decoding — a more complicated situation arises than during transmitting a normal signal. More operations have to be done in the same time unit, because the transformation of system A into system B and B into A system takes place.

Here the information (message) might be interfered, disrupted and even altered. But usually with one defective phoneme the alternation of information is not a serious one since auditory perception of language seems to be based on discrimination not of a single phoneme, but of a series of phonemes in prescribed order;[7] then, because speech cues are highly redundant so the child must establish the appropriate meaning from the context, which often extends beyond the sentence. Usually language learning comprehends meaning from the total context of what is said.

The possibility of an impaired information and disrupted communication is greater and more probable, however, in case where there are more defective phonemes in the individual's language system, in other words, with the greater frequency of the disrupted signals.

[7] M. F. Berry, *Language Disorders of Children* (New York, 1969), p. 123.

If, for instance, a Slovak child pronounces identically the sounds [s] and [ʃ], [z] and [ʒ], [ts] and [tʃ], [dz] and [dʒ], then in both his motor and acoustic area of the CNS has one pattern only for a couple of phonemes. He does not make any distinction between the words such as: [kosa — koʃa, koza — koʒa] etc.

In such cases (in transforming system B into system A and vice versa) operations are overloaded and the disturbance of information and communication is more probable. Also the consequences arising out of not understanding and misunderstanding are more disagreeable.

Let me present the case of an 11 years old Slovak girl, mentally quite normal, whose language system has 12 phonemes instead of 43 phonemes in the system of Standard Slovak. Besides 4 vowels /i/, /u/, /a/, /o/, consonant phonemes are represented by 3 labials /p/, /m/, /w/, 4 (interdental) palatals /t,/, /n,/, /l,/, /j/ and 1 laryngeal /h/. In fact the situation is still more complicated as consonant sequences are simplified and in some positions consonants are ommitted.

The palatal interdental /t,/ is the most frequent phoneme in the system B standing for 14 phonemes of the system A /t/, /d/, /t,/, /d,/, /s/, /z/, /ʃ/, /ʒ/, /ts/, /dz/, /tʃ/, /dʒ/, /k/, /g/.

The more dispersion a phoneme has the smaller is its distinctive capability, the less degree of information it provides.

Most of the distinctive features of the system A are absent in the system B, such as: voiceless/voiced /t/ — /d/, diffusive/compact /t/ — /t,/, interrupted/continuant /t/ — /s/, mellow/strident /t/ — /ts/, sharp/plain /t/ — /k/. For instance the word [t,ot,a] in B system stands for 392 possible words of the A system (e.g. [kosa, koza, koʃa, kotʃa, d:oza, so:da, sot,ia, zotʃia] etc.).

The capacity of the information in the word [t,ot,a] from the source B conveys very little — practically nothing to the receiver A.

It may be said that speech codes on the level of a phoneme or a word are individual and not corresponding with codes of system A. And that is why decoding of transmitted information from the source B by the receiver A is very difficult, or even impossible.

In cases where the receiver A fails to decode B's information (if he does not understand B's message), he asks the information to be repeated. After failing to understand it again, he might require the source B to clarify the message. During several unsuccessful attempts to transmit an intelligible information a great number of operation takes place in B's CNS which makes him exhausted and inervated much sooner than is usual with the normal individual of the A system.

A tension arises inside the source B which leads either to passivity and resignation or to irritation and agressiveness. It might lead to the state of mind which is called frustration in psychology.

The tension may be raised to greater power by the tension between the source B and the receiver A. And because the source B and the receiver A are human beings, there might arise various situations in the relationship between them. The manner in which the receiver A receives the information from the source B and the way how he responds to it has direct and substantial influence on the psychical state of the source B and on the further course of communication as well. Let us mention some of them.

(I) If the receiver A does not show to the source B that he has noticed his articulation disorder, communication might go on smoothly, undisturbed.

(II) But it is more likely that the receiver A when decoding distorted information from the source B shows he has noticed disrupted code. He then may react in two ways:

(1) The receiver A receives a distorted code and the tension might arise in him, but his next reaction is normal again (as he had not noticed the articulation disorder). But unfortunately this case is not very often.

(2) Far more frequent is the case when receiver A registers a distorted code from source B and indicates to source B that the code does not correspond with the code of receiver A. He indicates that it is a disturbing one, that his signal is not an explicit or intelligible one. He may ask source B to repeat his information, or he may even reject it and laugh at the speaker.

Such unpleasant possibilities must make us realize that articulation therapy is not only the teaching of the correct production of individual sounds, but also inevitably includes the rehabilitation of the whole personality, his psychical state and character, as well as his social adjustment. And last but not least, it also includes teaching the community through education how to keep in communication with individuals who have articulatory disorder.

DISCUSSION:

KACZMAREK: Es gibt einen Unterschied zwischen dem phonetischen und phonologischen System, weil das Kind sein phonologisches System nicht auf Grund seiner eigener Aussprache, sondern auf Grund der Aussprache seiner Umgebung aufbaut. Deshalb, wenn das Kind Ausssprachestörungen hat, dann hat es normales phonologisches System. Sollte das phonologische System gestört werden, dann haben wir es nicht mit Aussprachestörungen zu tun, sondern mit Sprachstörungen. Jedenfalls ist das Problem sehr interessant, muss aber noch veiter erforscht werden. (Siehe auch meinen Beitrag: "Über die Realisierung des phonologischen Systems durch Dyslaliker".)

PAČESOVÁ: Is there any difference in mastering the sibilant phonemes in Slovak- and English-speaking children?

SÖDERBERGH: What prof. Liška has told us about the effect of criticism on the psychical state of the defective speaker might have bearing on normal children acquiring the adult phonemic system from c.1-5 years of age. If you correct too much you might frustrate, if on the contrary you give the children good patterns by talking much to them they might become 'competent speakers' earlier and more easily.

ANSWERS:

To prof. Kaczmarek: I believe a lot of investigation and research has to be done yet in this field. On the whole I feel I have made distinction between the phonetic and phonemic (phonological) changes in articulatory disorders. I do agree with prof. Kaczmarek that only the change in the phonological system means a language disorder in children.

To dr. Pačesová: It does not matter whether the English /θ/ is changed into /s/ or /s/ is pronounced as /θ/. In case these two phonemes are realized identically, it means the change of the phonological system in English. I think of the stage in which some English children have one phoneme instead of two.

To dr. Söderbergh: The best way how to 'correct' a child's misarticulation in the stage he is acquiring speech is to present a good (correct) pattern of pronunciation to him. Of course, this depends on the emotional life of children. Some of them ask to be corrected, but most of them might be discouraged when corrected very often.

THE SYNCHRONY OF DIFFERENT DEVELOPMENTAL STAGES IN THE LANGUAGE OF A TWO-YEAR-OLD CHILD

KLÁRA S. MEGGYES

(Budapest)

Language is — as stated by Roman Jakobson — of stratified structure, within every level, beginning with the smallest unit, the phoneme, to the highest speech units.[1] And the child language is, as it is widely known, one of the most suitable domains to investigate the development, the build up of this stratification.[2]

Even in its unfinished stratification the child language forms a consistent system in each development stage. The system and its elements alter in the subsequent development periods. Henri Wallon's view about development in intelligent action also applies to the development of language. Thus, this development is "a series of transitions from one system to the other. As the structure of the system is differing, there is no such development result that itself does not change during transition."[3]

The present paper deals with the phenomena of child language, showing the simultaneous presence of characteristic features of different development periods, at a given stage. In other words: the object of the investigation are the recognizable elements of the change in a given synchrony, the main point of view being however the simultaneity and not the development or the change. This

1 Roman Jakobson, *Child Language, Aphasia and Phonological Universals* (The Hague-Paris, Mouton, 1968), p. 92.
2 See at the level of phonology: Jakobson, 1968: 92.
3 Henri Wallon, *L'évolution psychologique de l'enfant* (Paris, 1941). [In Hungarian: *A gyermek lelki fejlődése* (Budapest 1958), p. 18].

contribution intends to supply some data to the investigation of dynamism within synchrony. The dynamic character of a synchronous system has also been stated before (may I refer again to Prof. R. Jakobson),[4] though everybody engaged in language as a living organism has to rediscover this truth. Child language lends itself specifically to the investigation of this problem. The detailed analysis of the corpus exceeds the limits of the presentation, only some characteristic examples can be discussed here. The examples are taken from the language of my daughter Martha at the age of approximately two years.

(a) The best known kinds of manifestation for synchronous presence of two subsequent development stages are the cases in which parallel to the more or less developed new phenomenon, the already exceeded phenomenon lives also — unchanged. Such is the fact — observed by many authors — that babbling persists after the appearance of intelligent speech and also mispronunciation from time to time of already correctly pronounced sounds should be mentioned here. These are phenomena typically characteristic for transition periods. Further on a similar phenomenon may be observed in the formation of sentences as well. In the language of my daughter at the age of 1 year 8 months and a half several one-word sentences, put beside each other forming unshaped utterances, alternated in a special way with shaped sentences of two members (or two elements). (The latter were of verbal predicate + noun with the accusative suffix used as an object, though the definite article before the object was still missing. The lack of the personal pronoun is not a peculiarity of child language but a characteristic feature of the Hungarian language in general, the category of person being expressed by the personal suffixes of the verbs.) Examples for shaped two-member sentences: *Mekkész cukojt* (*Kérek még cukrot* 'Some more sugar please'); *Kéjek kockát* (*Kérek egy kockát* or *Kérem a kockát* 'Give me the dice please'); *Mekkapod kanalat* (*Megkapom a kanalat* 'I get the spoon'). Of

[4] For example: R. Jakobson, *Hang — Jel — Vers* (Budapest, 1969), pp. 166-67. [In original: "Pattern in linguistics: Contribution to Debates with Anthropologists".]

the set of one-word sentences consisting of unshaped utterances, a single example is cited: *Ënkentöje. Szobába. Nószi. Szobába. Hosszomka.* (*Zsebkendője. A szobába*[*n*]. *Nózi* [*orrocska*]. *A szobába*[*n*]. *Hozom vagy: hozd?* 'Her handkerchief. In the room. Nose. In the room. I'm bringing it or: bring it.') (The child asked for her handkerchief left in the other room.) It may be observed that part-units of the unshaped utterance i.e. word sentences are shaped even here and are provided with the adequate suffixes, however the whole is merely an amorph formation and should be considered as a more primitive extension of the one-word sentence from the point of view of formation than a regular two-member sentence. (At the same time the more primitive form might be, of course, an enrichment with regard to expression.)

Similar to this is a peculiarity related to word learning and word forming, i.e. to the aspect of meaning of the vocabulary. Words (nouns) created by name transfer are characteristic in general for the stage of one-word sentences and even here rather for its earlier stages. Some products of these periods are still living at a later stage, even new ones appear occasionally. — In Martha's case it could be observed at the stage of shaped sentences of two, three, and even more members. In general the naming of books, toys and other articles of use, actually based on misunderstanding belong to this stratum of vocabulary. For instance: *kukuri* is the name of the first picturebook, because of the cock on the front-page (*kukuri* is derived from the word *kukurikú*, imitating the voice of the cock, which is the cock's name in her language, learned from adults, an element of so-called nursery language). Or: *frizura* (*fizula, szizula* 'hair-do' as the name of the comb), etc. These words co-exist with words, the meaning structure of which — only roughly of course — agrees with their meaning in adult language. Most part of her vocabulary consists of the latter group, there are few name transfer words, this kind of word forming itself hardly exists, though it is still effective.

(b) Another possibility of co-existing results of two different development stages, or more exactly of the results of the already surpassed stage and of the new period is observed, when the old element is still living in its unchanged form, however there starts some kind of fitting into grammatical structure. So for instance words having become permanent in a single grammatical form are used by the child further in a single form. Such is the element *monádédit* of the sentence *Monádédit kérsz* (*limonádét kérek* 'some lemon squash please'), used at the age of 1,6-1,7 years, which was applied by her in freely constructed sentences also at the age of approximately two years: *Kérem*! *Kérem*! *Monádédit*! (1;9,22), *Mékkész monádédit*! (1;9,22) but always in a function suitable to the grammatical form (the quoted sentences are actually later variants of the above sentence used earlier in a single form). Thus in such cases the 'static' single-form words are fitted into the developing or developed inflection system, though their form remains further constant. Both vocabulary and morphological system also show a particular stratification in this respect: the dynamic elements and those fitted in as unified blocks, live parallel side by side and though the use of the latter is restricted, it is not contradictory to the rules applying to dynamic elements.

Similar observations can be made with regard to the conjugation of verbal forms used in the function of the first person singular. For my daughter, at the beginning, the second person singular verbal forms performed the above-mentioned function, based on direct imitation, alternating from time to time with verbal forms for the third person and the first person. Within the observation period a rather striking change took place in this respect; the use shifted to the third person singular verbal form used with the personal name (e.g. *Öltözik a Márti* 'Márti is dressing', 1;11,20). (In general this is the most frequent verbal form in the above function of the child language.)[5] The importance of the second person singular forms steadily diminished, in fact near the end of the observed period one verb alone occurred in this form, the fused *mékkérsz* (*Még kérsz* = *még kérek* 'some more please'), being at this time already such a static block as among nouns the

[5] Regarding the Hungarian see: e.g. J. Balassa, *A nyelvtani alakok kialakulása a gyermeknyelvben* (Magyar Nyelvör XLIX, 1920), p. 58; E. Kenyeres, *A gyermek első szavai és a szófajok föllépése* (Budapest, 1926), pp. 32-33, etc.

accusative form *monádédit*. However its fitting into the inflection system was less unambiguous, as there was no uniformity in the use of the personal suffixes of verbs.

(c) Another form of persisting language phenomena exceeded in development, is the type, when the old reappears only in certain defined situations, and does not occur otherwise any more. Such is the observation of Mrs. Ruth Hirsch Weir that babbling reappears from time to time in the monologues of her two and a half year old boy before his falling asleep.[6] Of my own observations I cite an example with relation to the stock of sounds. To this purpose some data have to be mentioned first. In the given period Martha was familiar with nearly the complete phoneme stock of the Hungarian language (the *r* sound developed only at the end of the observed period). Chiefly in the pronunciation of consonants, however, there was much fluctuation, nearly all had several variants (with the exception of the three voiceless stops, the *p*, *t* and *k* as well as the nasal *n*). The number of variants showed in general a decreasing tendency, proportional to the corresponding degree of stabilization in the sequence of consonant appearance, especially described by Jakobson. The voiceless pronunciation of the voiced sounds was the most conspicuous and most frequent among the variants. With other words they were substituted by their voiceless counterparts, especially in case of stops and fricatives. However the sound *b* was an exception: no voiceless pronunciation occurred in her speech (with the exception of the regular assimilation, of course). In verse or more exactly in the words of a song, — belonging lingually to the most conservative speech elements — the voiceless *p* occurred instead of *b* in constrast to the communicative speech. Partly because she did not understand the text. For instance: "Télapó it nan Ó a *supája* É a cipője" (1;10,21). (*Supája* instead of *subája*. 'Santa Claus is here, his coat is made of snow, his shoes of ice'.)

(d) Another characteristic manifestation of co-occurrence of the old and new stages of development is a kind of variances in sentence

6 Ruth Hirsch Weir, *Language in the Crib* (The Hague, Mouton, 1962), p. 23.

repetitions, being frequent in the language of the observed child. One of the components of the trying — variating manner of speech, ascribable to several causes, is an intention to correct: she immediately corrects the defectively pronounced sentences. Such are: *Ēbiszi nénikét. Ēbiszi néni.* (*Elviszi a nénit. Elviszi a néni, ti. a cicát* 'Auntie is taken away. Auntie takes away, i.e. the cat') (1;9,22). Or: *Adide teluzát a Máti! Adoda teluzát a Mátikának!* (*Add ide M-nak = nekem a ceruzát.* 'Give Márti = me the pencil!') (1;10,21). This repetition with the purpose to correct is generally frequent in child language. It is usually observed in the correction of defectively pronounced sounds, but as shown also by above examples it occurs too on the level of the paradigm lines or the sentence.

(e) Finally the most interesting manifestation of the synchronous presence of phenomena, belonging to different development stages, should also be mentioned, which is in closest connection with the general problems of the whole child language development. Namely how the developing new elements lean on the already existing, established ones. This phenomenon is manifested in a most demonstrative way in the developmental sequence of the parts of sentence. As an example the development of the dative is quoted in the language of my daughter: in the stabilization of it the noun without suffix (nominative) and chiefly the accusative had the role of a substitute, though the latter has no formal relation to the Hungarian dative. This problem could merely be mentioned here, it is dealt with in detail somewhere else and will be followed up subsequently.[7]

From the above mentioned facts it may be concluded that the enumerated appearance manners are not tied to certain language levels. The same kind of phenomenon may occur on the level of phonology, of vocabulary, of morphology and of syntax alike, even though perhaps not exactly at the same time. Each separate domain of the language develops by itself if not independently

[7] See in detail K. S. Meggyes, *Egy kétéves gyermek nyelvi rendszere* [The Language of a Two-Year-Old Child]; pp. 50-51, 70-71. (Published in 1971, Budapest, Akadémiai Kiado.)

from other domains, however qualitative changes are not necessarily simultaneous. Therefore it may be possible that a given state of language is on one level in a new stage of development, whereas on the other level the new stage is just beginning or has not even started yet. This phenomenon still further colours the problem of complicated relations of movement and state in language and gives quite a particular synchrony in the development of child language.

DISCUSSION:

DEZSŐ: I touched upon the question of development in child language. Considering e.g. three developmental stages (stage 1, 2, 3) we can find phenomena in stage one which may be considered as relevant in stage 2, but beside them, phenomena relevant only for stage 3 may also be present. This was illustrated by an utterance consisting of word sentences from Meggyes' paper.

SENTENCE-PROGRAMMING SPAN IN CHILD LANGUAGE

M. MIKEŠ, L. DEZSŐ, P. VLAHOVIĆ

(Novi Sad, Budapest)

The aim of this paper is to propose universal principles of sentence-programming span in child language, based on the analysis of the speech material of four children in the period from the first appearance of two-word utterances till the pre-transformational stage of compound sentences and complex sentences with an embedded transform. The developmental phase of an utterance is determined by the number and complexity of main operations its sentence-programming rule contains. We are analysing minimal contextual units which may, but need not, be sentences, because such contextual units as sentence blocks appear to be relevant for the sentence-programming span in child language.

The analysed speech material was compiled by longitudinal observations of two bilingual children — D 1, born in 1958, D 2, born in 1961 —, who were exposed to the linguistic influence of both Hungarian and Serbocroatian environment, and two monolingual children — D 3, born in 1962, D 4, born in 1965 —, who were exposed to the linguistic influence of Serbocroatian environment.

Sentence programming follows rules which include two types of syntactical operations:

(1) universal syntactical operations

(2) language restricted syntactical operations.

We are going to treat only the first ones, because our intention is to propose universals. Some language restricted operations, such as morphological devices, the use of prepositions and postpostions, as morphological devices, the use of prepositions and post-positions, etc. have been presented and discussed in our previous studies (1966b; 1967). The generative use of the word order, which also belongs to this type of operations has been touched upon in Slobin's (1969) an our works (1970; in press).

Universal syntactical operations do not depend on the language system the child is acquiring. They represent the minimum of operations needed for the generation of a contextual unit, and work on two basic principles: adding of components and conjoining of components. (See Table I.)

TABLE I

Developmental phase in SPS	Minimal number of components	Number of operations	The new operation	
			Conjoining	Adding
1. Word sentence	1	1	+	
2. Word-sentence block	2	2		+
3. Simple sentence	2	3	+	
4. Simple-sentence block	3	4		+
5. Complex sentence	3	5	+	
6. Complex-sentence block	4	6		+
7. Compound sentence[a]	4	7	+	
8. Compound-sentence block	5	8		+

[a] Or complex sentence with an embedded transform.

A WORD SENTENCE consists of only one morpheme, i.e. word, which stands for Pro Predicate (not an elliptic one-word sentence as it is known in the speech of adults).

(1) *titi* (D 2 1;2)

D 2 asks for water. (*tütü* H (*ch*) 'water')

A WORD-SENTENCE BLOCK consists of two or more word sentences forming a unity which may be transformed into a simple sentence. The word sentences are seperated by a disjointed intonation marked by a pause.

(2) *tütü # daj* (D 2 1;3)

D 2 asks for water. (*daj* SC 'give')

A SIMPLE SENTENCE contains two morphemes — words or endings —, one of which stands for Predicate and the other which may have the function of an Agent, Object, Experiencer, Locative or Instrumental. An ending is considered as a seperate morpheme only if it satisfies the conditions of our test which is discussed in our previous works (1966a; 1967), and is similar to the procedure of Blount (1969: 66-67).

(3) *apu tütü* (D 2 1;4)

D 2 is asking her father (*apu* H 'father') for some water.

A SIMPLE-SENTENCE BLOCK consists of at least one simple sentence and one or more word sentences. The simple sentence is separated from the word sentence or from the other simple sentence by a pause, but they all together represent a unit which may be transformed into a complex sentence or a compound sentence.

(4) *tütü nema # tütü ó* (D 2 1;5)

The child is stating *no water # water there* (*nema* SC '(there is) no', *ó* H (*ott*) 'there').

A COMPLEX SENTENCE contains at least three morphemes, one of which stands for the Pro Verb or Verb and the two other morphemes representing two other components of the sentence.

(5) *tátá nicse itt* (D 2 1;7)

The child is stating *bag isn't here*. (*táska* H 'bag', *nicse* H (*nincsen* 'isn't', *itt* H 'here')

A COMPLEX-SENTENCE BLOCK contains at least four morphemes. Three of them, forming a complex sentence, are separated from the fourth by a disjointed intonation, but they all together represent a unit which may be transformed into a compound sentence (6 i, 6 ii) or a complex sentence with an embedded transform (6 iii).

(6) i *apu menj innen # csücsüj le* (D 2 1,8)
'Father go away # sit down!'
ii *apu nincs mami # nincs mami eveszett* (D 2 1,8)
'Father, no mother # no mother (because) she got lost'.
iii *crtalam kucu # crtalam gavu* (D 2 1,10)
'I have drawn a dog # I have drawn the head'.

This is a pre-transformational form of *crtala sam kucinu glavu* ('I have drawn the head of the dog').

The last two developmental phases: COMPOUND SENTENCE and COMPOUND-SENTENCE BLOCK fall outside the examined period, so they are not discussed in this paper.

To this minimum of syntactical operations four additional syntactical operations may be taken into consideration in the sentence-programming span:

(1) addition of the operation of negation,
(2) integration of the operation of negation by transformation,
(3) addition of the operation of interrogation,
(4) integration of the operation of interrogation by transformation.

Ursula Bellugi's analysis of the development of the negative and interrogative utterances shows a succession of short-lived devices for performing grammatical transformations such as negation and interrogation (Slobin, 1967: 112-16). Velta Rūķe-Draviņa's observations on the structuring of negative sentences are also very instructive in this respect (1963: 47-48). It would be quite reasonable to suppose that the addition of negation or interrogation affected the sentence-programming span in child language. However, in the scope of the present paper we have no possibilities to discuss their effect in the speech material of our children.

The validity of the proposed developmental phases in sentence programming span has been checked by quantitative analysis of the compiled speech material, the results of which we are going to sum up.

D 1

As the evolution in sentence-programming span from the appearance of phase 2 to phase 6 in D 1 lasted just half the time

(3 months) that was needed in the case of D 2, D 3, and D 4, two surveys are made for this child, one representing the three-month period in three one-month stages and the other in six half-month stages.

The first month is marked by a predominance of phase 1 (word sentences), while the predominance of phase 3 (simple sentences) is obvious in the next two months. The difference between the second and third month is significant only for the increase of phase 5 (complex sentences) from 7% to 12%, as well as for the first appearance of phase 6 (complex-sentence blocks) and 7 (compound sentences or complex sentences with an embedded transform).

When analysed in six half-month stages the development of sentence-programming span of this child does not differ essentially — as we are going to see — from that of the other three children whose development comprises a six-month period. We may divide it into two major developmental stages, the first comprising the first three half-month stages, marked by the presence of phases 1 to 3, and the second, comprising the other three half-month stages marked by the appearance of phase 4 to 7 and the complete disappearance of phase 2 in the fifth developmental stage. On the whole, the transitory phases — 2,4 and 6 — seem to be of lesser importance in the sentence-programming development of this child.

D 2

The evolution of sentence-programming span in D 2 proceeds in a smooth way. If the quantitative tendencies in phase 1 are observed, it may be concluded that the percentage amount decreases constantly and gradually from 78% at 1;2 to 35% at 1;7. The decrease of phase 2, however, shows a stoppage — a slight growth from 15% at 1;3 to 19% at 1;4 and a stagnation at 13% in 1;5 and 1;6.

The appearance of phase 3 is noticed at a rather early stage (1;3) and in a relatively high percetage (20%), but it does not make any significant progress until 1;7 (36%).

The development of phase 4 and 5 is characterized by a slow growth from 1;4 to 1;6 (from 2% at 1;4 to 4% at 1;6) and a rather important increase at 1;7 (from 4% to 12% in phase 4 and 3% to 8% in phase 5).

D 3

In the course of the six-month period we may distinguish three developmental stages, each lasting two months:

(1) 1;2 and 1;3 are marked by the predominance of word sentences and a slight decrease of the quantity of word-sentence blocks.

(2) 1;4 and 1;5 are characterized by the appearance of simple sentences. Both in word-sentence blocks and in simple sentences a tendency of decrease in quantity is to be noticed.

(3) 1;6 and 1;7 aıe the most dynamic two months, marked by a sudden decrease of the quantity of word sentences and by a very rapid widening of the sentence-programming span by the appearance of simple-sentence blocks, complex sentences, complex-sentence blocks and even some compound sentences at the end of the examined period.

Phase 2 is constantly decreasing and almost disappearing in the last two months (only 1%). In the development of the simple sentences we may notice a falling back in the first two months of its development.

D 4

The first half of the six-month period (1;3 to 1;5) is marked by a narrow sentence-programming span. The great majority of utterances is generated by a simple operation (word sentences), and a small percentage by two operations (word-sentence blocks).

In the second half of this period (1;6 to 1;8) we have noticed a dynamic evolution in sentence programming. The percentage of word sentences decreases from 74% at 1;6 to 41% at 1;8. The quantity of word-sentence blocks increases a little at 1;6, but in the following two months it decreases rapidly to only 1% at 1;8. The greatest progress is noticed in programming simple sentences. At 1;8 they represent 36% of all utterances, which is nine times more than at 1;6 when simple sentences appear for the first time. As to the quantity of simple-sentence blocks and their slight dynamism in the evolution of sentence programming, we may compare phase 4 to phase 2. The first complex sentences appear at 1;7 and they make rather remarkable progress until the end of the examined period, increasing from 3% to 12%.

In recent research work on child-language syntax the investigators (Slobin, Ervin-Tripp, Miller, Bellugi, Menyuk, Braine, Kernan, Klima, Bloom, Brown and Fraser) have paid attention to the 'two-word' or 'two-morpheme' utterances. In our terminology they comprise two types of structures: word-sentence blocks and simple sentences. Unlike the investigators of classical linguistic formation, who considered child language as a more or less successful imitation of adult speech, these authors assume that the combinations of two words were formed on the basis of productive rules.

We add another presumption: There is a paralelism between the operational principles in syntax and phonology at the developmental stage which represents the transition from the phonological structure C 1 + V 1 + C 1 + V 1 to the phonological structure C 1 + V 1 + C 2 + V 1 or C 1 + V 1 + C 1 + V 2 in the development of the sound system, and the stage which represents the transition from the phase of word sentences to the phase of word-sentence blocks and simple sentences in sentence-programming span:

(7) i *bu-bu* (D 4 1;3)
(*bu-bu* > *ljubi* SC 'she kisses')
ii *juka* # *juka* (D 4 1;5)
(*juka* > *lutka* SC 'doll')
iii *juka* # *baja* (D 4 1;6)
(*baja* SC/ch/ 'she sleeps')
iv *juka baja* (D 4 1;7)
('The doll sleeps')

On both levels — phonological and syntactical — the structuring is based on the repetition of one entity — syllable, in the case of 7 i and word sentence, in the case of 7 ii. The repetition in 7 i is the starting point in the development of the phonological structure *ljubi:*

bu-bu > **bubu* > **bubi* > *ljubi*

In the same way the repetition in 7 ii is the starting developmental point of the syntactic structure 7 iv.

The two other transitional phases in sentence programming are the phase 4 and 6. Of special interest for further development would be the analysis of the phase 6, which may be regarded as a "pre-embedding" phase (Slobin, 1967: 120, Menyuk, 1969: 39).

Along with the child's ability to generate sentences by his creative intuition, the imitative ability of the child also plays an important part in sentence-programming. So, for instance, when analysing the speech development of D 3, we must take into consideration that this child is the first-born child in the family, always surrounded by adults, whose conversation presented a great deal of stimuli and patterns for imitation. It seems that D 3 acquired the language much more by imitation of the speech of her environment than our other children did. This fact may explain the 'jumping' character in the development of sentence-programming span.

When comparing the results obtained for the bilingual children, D 1 and D 2, to those of the monolingual children, D 3 and D 4, no relevant difference between the bilinguals and monolinguals can be stated in the evolution of sentence-programming span, because the individual differences between D 1 and D 2 or between D 3 and D 4 are more obvious than the differences observed between D 1 and D 3 or between D 2 and D 4.

However, some instructive hints may be given with respect to the differences between bilingual and monolingual children:

(1) In monolinguals a stagnation interval can last two to three months, while in bilinguals it lasts only one to two months.

(2) An earlier appearance of both simple and complex sentences is noticed in bilinguals.

(3) The amount of complex sentences may be considerably larger in monolinguals, as is shown in the case of D 3.

On the ground of the analysis done we are going to give some conclusions which may be proposed as universal principles to be checked on speech material of other children:

(1) The succession of phases in the sentence-programming span of all four children corresponds to out theoretical presumption of the succession of phases, which has been demonstrated.

(2) The quicker the speech development, the lesser the importance of transitional phases which represent the pre-transformational stages in sentence-programming span.

(3) The transitory phases always show a decreasing tendency due to the constant growth of the adaptive ability of the child to his linguistic environment.

(4) The increase of the quantity of simple sentences is conditioned by the decrease in the amount of word-sentence blocks, although some stagnations are observed, especially in the first half of the examined period.

(5) Simple-sentence blocks which represent the second tiansitional phase, both in frequency of their appearance and in their duration, are of lesser importance for the development of sentence-programming span than the word-sentence blocks, the first transitional phase.

REFERENCES

Blount, B. G.,
1969 *Acquisition of Language by Luo Children* (= *Working Paper* 19, Language-Behavior Research Laboratory, University of California). (Berkeley).

Dezsö, L.,
1970 "A gyermeknyelv mondattani vizsgálatának elméleti-módszertani kérdései", *Általános Nyelvészeti Tanulmányok* VII (Budapest).

Dezsö, L., and P. Vlahović,
In press "Teorijsko-metodološka pitanja izučavanja sintakse dečjeg govora", *Prilozi proučavanju jezika* 6 (Novi Sad).

Kernan, K. T.,
1969 *The Acquisition of Language by Samoan Children* (= *Working Paper* 21, Language-Behavior Research Laboratory, University of California) (Berkeley).

Menyuk, P.,
1969 *Sentences Children Use* (= *Research Monograph* 52) (Cambridge, Mass.).

Mikeš, M., and P. Vlahović,
1966a "Metoda analize sintaksi: čkih struktura u dečjem govoru", *Prilozi proučavanju jezika* 2 (Novi Sad).
1966b "Razvoj gramatičkih kategorija u dečjem govoru", *Prilozi proučavanju jezika* 2 (Novi Sad).
1967a "Acquisition des catégories grammaticales dans le langage de l'enfant" *Enfance* 3-4 (Paris).
1967b "Approche méthodologique aux recherches de l'apprentissage du langage par l'enfant", *Actes du Xe Congrès International des Linguistes* (Bucarest).
1968 "Acquisition du système associatif des sons", *Bulletin de la Société de Linguistique* (Paris).
Rūķe-Draviņa, V.,
1963 *Zur Sprachentwicklung bei Kleinkindern*, 1: *Syntax* (Lund).
Slobin, D. I.,
1967 *A Field Manual for Cross-Cultural Study of the Acquisition of Communicative Competence* (University of California, Berkeley).
1969 *Universals of Grammatical Development in Children* (= *Working Paper* 22, Language-Behavior Research Laboratory, University of California) (Berkeley).

APPENDIX

SPS Statistical survey D 1 (A)

The survey comprises a three-month period. The beginning of it (1;2) is marked by the appearance of phase 2 of SP. At the end of the period (1;4) the presence of phases 6 and 7 is noticed.

Age	Phase 1	Phase 2	Phase 3	Phase 4	Phase 5	Phase 6	Phase 7	Total
1;2	52	3	20					75
	69%	4%	27%					
1;3	67	3	91	8	13			182
	37%	2%	50%	4%	7%			
1;4	119		206	17	49	11	5	407
	29%		50%	5%	12%	3%	1%	

SPS statistical survey D 1 (B)

As the evolution in SP from the appearance of phase 2 to phase 6 in D 1 lasted just half the time that was needed in D 2, D 3 and

D 4, we have divided the three-month period into six half-month stages.

Age	Phase 1	Phase 2	Phase 3	Phase 4	Phase 5	Phase 6	Phase 7	Total
1;2 (I)	21	2	3					26
	81%	8%	11%					
1;2 (II)	31	1	17					49
	63%	2%	35%					
1;3 (I)	25	1	12					38
	66%	3%	31%					
1;3 (II)	42	2	79	8	13			144
	29%	1%	55%	6%	9%			
1;4 (I)	74		104	13	21	7		219
	34%		47%	6%	10%	3%		
1;4 (II)	45		102	4	28	4	5	188
	24%		54%	2%	15%	2%	3%	

SPS statistical survey D 2

The survey comprises a six-month period. The beginning of it (1;2) is marked by the appearance of phase 2 of SP. At the end of the period (1;7) the presence of phase 6 is noticed.

Age	Phase 1	Phase 2	Phase 3	Phase 4	Phase 5	Phase 6	Total
1;2	42	12					54
	78%	22%					
1;3	58	13	18				89
	65%	15%	20%				
1;4	84	25	20	2	1		132
	63%	19%	15%	2%	1%		
1;5	86	22	31	7	5		151
	58%	13%	21%	5%	3%		
1;6	149	34	63	10	8	1	265
	57%	13%	23%	4%	3%	0,4%	
1;7	162	30	165	54	36	11	458
	35%	7%	36%	12%	8%	2%	

SPS statistical survey D 3

The survey comprises a six-month period. The beginning of it (1;2) is marked by the appearance of phase 2 of SP. At the end of the period (1;7) the presence of phases 6 and 7 is noticed.

Age	Phase 1	Phase 2	Phase 3	Phase 4	Phase 5	Phase 6	Phase 7	Total
1;2	5	2						7
	71%	29%						
1;3	10	2						12
	83%	17%						
1;4	12	3	4					19
	63%	16%	21%					
1;5	37	4	8					49
	76%	8%	16%					
1;6	32	1	21	2	24			80
	40%	1%	26%	3%	30%			
1;7	54	3	86	6	51	5	7	212
	25%	1%	41%	3%	24%	3%	3%	

SPS statistical survey D 4

The survey comprises a six-month period. The beginning of it (1;3) is marked by the appearance of phase 2 of SP. At the end of the period (1;8) the presence of phase 6 of SP is noticed.

Age	Phase 1	Phase 2	Phase 3	Phase 4	Phase 5	Phase 6	Total
1;3	12	2					14
	86%	14%					
1;4	19	2					21
	90%	10%					
1;5	21	3					24
	88%	12%					
1;6	35	9	2	1			47
	74%	19%	4%	3%			
1;7	80	13	30	11	4		138
	58%	9%	22%	8%	3%		
1;8	46	1	41	10	14	1	113
	41%	1%	36%	9%	12%	1%	

DISCUSSION:

KAPER: Aus den angeführten Beispielen geht m.E. hervor, daß das Kind nicht zwei, sondern nur EIN System besitzt; es ist bloß nicht imstande, die phonematischen Unterschiede zum Ausdruck zu bringen.

PAVLOVIĆ: Deux communications de Mme Mikeš, M. Dezsö, Mme Vlahović présentent les observations concernant l'apprentissage du serbe et de la langue hongroise, aussi les faits de bilinguisme (cf. la thèse de Mme Mikeš). Ce qui est à noter spécialement, c'est le caractère collectif d'observation et d'interprétation, qui est très précieux pour les conclusions générales.

A STUDY OF CHILDREN'S SPEECH IN A SPONTANEOUS AND NON-SPONTANEOUS SITUATION

NICHOLAS MILAN

(Bratislava)*

GENERAL INFORMATION

One usually begins with the most essential factors in teaching foreign languages. These factors usually consist of repeating expressions, words, grammar and drilling. We wish to point to broader connections of a language regarding the communicative medium in exchanging information. These connections have a meaning even in learning and teaching languages.

We tend to think of language exclusively as a communication device, but it is more than that. If it were not, we should all be speaking with the same tongue. Looking back through supertribal history it is easy to see how the anti-communication function of language has been almost as important as its communication function. More than any social custom it has set up enormous inter group barriers. More than anything else, it has identified an individual as a member of a particular super-tribe, and put obstacles in the way of his defecting to another group. (Desmond Morris, 1969).

Each complexity of a foreign language is a factor aimed against the foreigner. It serves to betray and trace the foreigner. This is known by the intelligence service which began to work on a safer system of gaining agents from natives of foreign nations due to unsuccesses with natives of their own nation who had learned a foreign language (second language). The factors are less evident in

* Pedagogical Research Institute in Bratislava.

learning a foreign language from the standpoint of enrichening knowledge than the component drilling of a language. These factors are sociability, group-attractiveness, willingness to get into contact with works, culture, persons speaking a foreign language, motivation according to the aims of learning a foreign language.

The sociability factor — the willingness to get into contact with the foreign language culture is rather hidden to the pedagogues. However, its effectiveness can be noted from the consequences. I would call such factors unconscious or preconscious (as you like) factors as we come to know them according to results although they are frequently not even known consciously to people effected by them. I am conscious of the unpopularity of this vague expression. However, I think as far as we do not get acquainted with it and as far as we shall not know how to use it currently, we shall not get further as regards the conscious of self-knowing. Even our forefathers did not consciously form a real linguistic obstacle in communication which is represented by a foreign language. This occurred as an advantage in the competition of super-tribes.

Children at the age of 2-5 years are very inventive in forming words (Čukovskij, 1959). This forming of words is done unknowingly. In the unconsciousness of children there is a usual accurate classification of grammatical forms of categories and columns, prefixes and suffixes which would, "mean quite enough difficulties even for grown ups. This classification seemed to me all the more miraculous that the child itself had no idea of it" (writer Čukovskij, 1959) ... "the child who has no idea of grammar will tell you quite correctly all the modes, genders, grammatical forms, words although it has no knowledge of these" (*ibidem*). "These recipes, or formulae, the child somehow discovers by intuitive processes — probably not unlike the unconscious inferences which go into scientific discovery — by the time it has reached the age of four" (Koestler, 1967). "The fact that we can freely produce sentences we had never heard before is amazing. The fact we can understand them when produced is nothing short of miraculous. A child never has a look at the machinery that produces English sentences. He could never have a look at that machinery. Nor is he being told about it since most speakers are completely unaware of it" (J. Jenkins, cited in Koestler, 1967).

In practical teaching of a foreign language, we cannot overcome the obstacle which forms a model situation against a spontaneous situation which represents a natural situation in life for the social and talkative foreigner. The difference is also an unconscious factor. Teaching therefore has different conditions from those in life itself and these are limited. The aims and method should be chosen according to this reality.

We are making experiments since 1962/63 at the Pedagogical Research Institute and introducing into the daily school practice some principles of programmed teaching. These principles represent raw material which can hardly be used in school without modification. Our experiments deal with teaching at the elementary school in grades 2-5, with children at the age of 8-11 years. We are making experiments in the fundamental teaching of the Russian language. Table I shows subjects for which we have developed whole year experimental programmed help textbooks.

TABLE I

SUBJECT (In the brackets is the name of author of programmed help textbook)	Grade	Number of pupils	Number of teachers
Slovak language (P. Bernáth)	2	8 000	266
Arithmetic (M. Barišová)	3	6 000	2 000
Arithmetic (M. Barišová)	4	3 000	1 000
Russian language (A. Koščová)	4 and 5	260	8
Slovak language (P. Bernáth M. Milan)	5	20 000	660
Arithmetic - Mathematics (B. Barišová, M. Milan)	5	270	9
Russian language (A. Koščová)	6 Hungarian	360	12
TOTAL		37 890	3 955

The fundamental teaching of the Russian language is programmed at two lessons weekly. The pupils are to learn how to read the printed and written forms of the Russian alphabet and gain a pretentious vocabulary in an active spoken and passive form. The most progressive pupils read tales and use the dictionary independently.

The text part of the programmed textbook is set in the tape recorder according to which the pupils learn the pronunciation and how to talk in the framework of the spoken part of the lesson. This is an exercise for the oral expression.

The pupils learn only at school. There is no homework. The whole programmed textbook according to which the pupils work remains at school. The pupils have no other textbooks. One lesson is divided into two parts. The spoken part about 15-20 minutes and the silent part from 25-30 minutes. In the silent part each pupil works independently on the basis of the programmed textbook. The textbook has three parts. The pupils get each part gradually according to how they finish self-pacing.

The pupils praise the spontaneous possibility of working without homework (from 296 answers 100; i.e. 33%, taking first place), they need not carry textbooks to school; silent independent work under the conduction but without interruption by the teacher etc.

RESULTS

Of the five basic principles of programmes (Fry, 1963), it was possible to use the following: (1) Small steps (without formality); (2) Active responding; (3) Self-pacing; (4) Improvement of programme, student testing. Immediate confirmation is a source of various difficulties in introducing the principles of programmes into schools.

The pupils work in such a manner that they continue in the next lesson where they have stopped in the previous lesson. The continuity in the lessons is preserved. The difference between the quickest and slow pupil is at a ratio of 1:3 to 1:4. The quickest pupils amounting to 37%, continue to read tales independently with the help of a dictionary during the silent part of the lesson. The teacher provides individual help to pupils if required by whispering so as not to interrupt pupils in their work. The teacher takes note of passive pupils and helps them individually. The third part of the textbook consists of tales.

The pupils in the first year learn 604 expressions and words besides reading and writing (the plan requires 150). Pupils who read tales exceed a vocabulary of 604 words.

Experimental textbooks are being tried out in the fourth grade of schools where teaching is done in the Slovak language and in the sixth grade in schools where teaching is done in the Hungarian language.

DISCUSSION

The research in the possibilities of making use of some principles of programmes in the teaching of languages at schools is a part of the State Investigation Plan 1966-1970 — "Increasing the Effectiveness of Communist Education". It seems that it is necessary to look for new ways in the conditions of teaching foreign languages at schools which would be appropriate to the possibilities of schools. Two lessons weekly means about 72 lessons yearly. This enables the pupils to learn about 8,3 new expressions in one lesson. Together it makes 604 expıessions which corresponds to 604 expressions besides learning how to read and write.

The limitation in learning at teaching in school (without homework) has a positive effect on the motivation of pupils. Knowledge is reinforced emotionally. The pupils develop a working endeavour during the lesson accompanied by physiological marks (blushing or getting pale).

When teaching by methods where programmed lessons are used, a technical solution is decisive. Lack of paper forces one to seek more thrifty methods. The application of transparent wrappers of polyvinilchloride is one solution. The programmed lessons and wrappers can be thus preserved. For writing an ordinary pencil is required and the wiping off can be done with a damp cloth and a spot of soap.

In the meantime we have not developed textbooks for further grades besides the first grade which would further spread knowledge. In spite of the fact that experimental teaching has been going on for five years, it is necessary to further verify results on examples of some ten thousands of pupils and at least a thousand teachers.

The method used under the conditions at schools probably mostly approaches the requirements of the spontaneous way of teaching. Further improvements are possible, i.e. by approaching the children with Russian expressions which they mostly use in the Slovak language as well as by improving the pedagogical tact of teachers in teaching the pupils. This means that it is especially necessary to correct the pronunciation and mistakes of pupils in the spoken language. In teaching foreign languages the words of E.H. Walters and E.B. Castel (1967), are valid: "Correction of Common Errors". "This seemed logical to many teachers: if children did not speak standard English they must be corrected and not allowed to speak their own version. The result of this attitude was that many children stopped speaking in school at all, because they were afraid of being wrong, and so they learned little."

REFERENCES

Čukovskij, K.,
1959 "Od dvoch do piatich" [From two to five, tr. from Russian], *Mladé letá* (Bratislava), 21; 22.

Fry, E. B.,
1963 *Teaching Machines and Programmed Instruction* (New York-Toronto-London, Mc Graw-Hill), p. 201.

Koestler, A.,
1967 *The Ghost in the Machine* (London, Hutchinson), pp. 29; 30.

Koščová, A.,
1967 "Programazott oktatás az oross nyelvben" [The Programmed Teaching in Russian Language], *Szocialiszta nevelés* [The Social Education] 3, 77.

1969 "Otázka koncepcie vyučovania ruskej morfológie a využitie metódy programovania" [The Concept Problem of the Russian language Morphology Teaching by Means of Programming], *Ruský jazyk* [The Russian Language] 2, 69.
1970a "Aké máme rezervy a efektivitu vo vyučovaní ruštiny?" [What Reserves and Effectivity Have We in Russian Language Teaching?], *Ruštinár* [The Russian Language], 1, 10.
1970b "Konštrukcia krokov a podávanie učiva v programovanej učebnici" [The Construction and Presentation of Lessons in Programmed Textbook], *Ruštinár* [The Russian Language] 2, 41.
Morris, D.,
1969 *The Human Zoo* (London, Jonathan Cape), p. 29.
Walters, E. H., and E. B. Castle,
1967 *Principles of Education, with Special References to Teaching in the Caribbean*, (London, Allen and Unwin), p. 134.

DISCUSSION:

LIŠKA: Do you expect better results in teaching foreign languages according to your programmed teaching? I think that rather unsatisfactory results in teaching foreign languages in our country are mainly because we start with foreign languages too late. The sooner a child starts with his second language the better for him. He can copy a good pattern of pronunciation, intonation and grammar without a drill. The fifth or sixth year of age is a good start.

PRŮCHA: (1) Do you have some proof of the result of the programmed teaching of Russian language on fourth grade?

(2) How do you know that the pupils really master 604 expressions and words?

ANSWER:

(1) We set many examinations to prove the results. One series was the comparison control — experimental classes. The tests were the same. Experimental classes were significantly better. Another series of testing was the teacher's opinion on the end of school year. Only one or two teachers from hundred did not consider that the pupils of experimental classes were better.

(2) The programmed textbook has 604 expressions and words. The traditional textbook has 150. By the end of school year almost all pupils had mastered the programmed textbook.

SLAMA-CAZACU: Do you envisage to extending the study to foreign-language teaching also, and for this teaching in the first grades?

It is an interesting paper, more pedagogical than psycholinguistic, dealing with the teaching of a foreign language. Have you also experiences with the teaching another language to young children?

ANSWER:

(1) Yes, the paper is pedagogical, deals with the beginning of foreign language instruction of children in the fourth grade, i.e. ten years old.

(2) With another foreign language we have no experiences. Only with the grammar in the mother tongue during the 2nd class, i.e. eight year old children.

LIŠKA: Do you compare the teaching of foreign languages in Czechoslovakia and abroad? Have they better methods than we have? For example in Norway they teach foreign languages from the 3rd grade, in elementary school.

ANSWER: I had compared the state of foreign language teaching in our country (Czechoslovakia) and abroad. But it is hardly comparable. We test long-time teaching by means of programmed textbooks during a whole year. It is not very extensive teaching. Represents only 72 lessons yearly, in most compelled school teaching two lessons weekly. You are right. We also try to push down the age level of foreign language teaching to the early ages. But the secretary of Education does not agree with this. Only in isolated cases we have foreign language instruction, for example in the kindergarten or nursery school.

UNE CONTRIBUTION À LA PÉDOPHONÉTIQUE COMPARÉE

KAREL OHNESORG

(Brno)

Lors de la simple comparaison des paroles enfantines du langage de plusieurs enfants de même milieu linguistique ainsi que par la juxtaposition des réalisations phonétiques dans le langage des enfants de divers groupes ethniques, on peut saisir beaucoup de ressemblances et de coïncidences. Plus une telle analyse est détaillée, plus on relève de similitudes dans l'évolution phonétique du langage enfantin.

De nos jours, on peut déjà se baser sur de riches données qui sont présentées dans une transcription phonétique satisfaisante. Il est alors possible de trouver une longue série d'exemples pour démontrer des phénomènes analogues ou identiques dans le développement phonétique du langage des enfants. Jusqu'ici, nous avons recueilli quelques traits communs du phonétisme enfantin dans le français, le letton, le polonais, le russe, le slovène et le tchèque et, passim, des documents allemands, anglais, italiens et roumains, en essayant d'en tirer quelques conclusions générales. Espérons que les documents seront, à l'avenir, encore augmentés par des exemples puisés dans le langage des autres enfants et que nos observations pourront être complétées et élargies.

Nous avons puisé nos documents tout d'abord dans les œuvres publiées par les auteurs suivants: Mme Suzanne Borel-Maisonny, Stefanya Brenstiern-Pfanhauser, Marcel Cohen, Kornej Čukovskij, Antoine Grégoire, A.N. Gvozdev, M.E. Chvatcev, Leon Kaczmarek, Rudolf Kolarić, Milivoie Pavlović, Mme Velta Rūķe

Draviņa, I.A. Sikorskij, Mme T. Slama-Cazacu, Paweł Smoczyński, Mme Maria Zarębina et les auteurs tchèques.

En attendant, nous avons constaté des ressemblances dans plusieurs phénomènes: la sonorité des consonnes, la différence entre le développement des occlusives et celui des fricatives, la réalisation tardive des consonnes vélaires, l'articulation des sibilantes, la formation de consonnes liquides, l'omission de la consonne initiale, les changements phonétiques dans le langage suivi, le raccourcissement des mots.

Pour le présent exposé nous allons analyser un seul de ces faits, l'omission de la consonne initiale dans les vocables de l'enfant. En étudiant des données dans divers ouvrages, on peut remarquer à plusieurs reprises ce fait intéressant et qui n'a pas été, à notre avis, jusqu'ici assez souligné par les autres auteurs.

Nous avons constaté ce phénomène dans le langage de plusieurs enfants tchèques. Cependant, on peut en trouver des preuves même dans les œuvres consacrées au langage des enfants étrangers. Nous laissons de côté les cas où l'enfant omet l'initiale parce qu'elle ne fait pas encore partie de son répertoire de sons, comme p.ex. s'il prononce *áva* au lieu de *káva*, ne sachant pas encore articuler la vélaire. Il s'agit ici des cas où les enfants quelquefois ne réalisent pas au commencement du mot la même consonne dont ils possèdent l'articulation depuis assez longtemps et qu'ils forment aisément dans les autres positions. Qu'il nous soit permis de citer quelques exemples de la langue des enfants de divers pays:

français: *osõ (chausson)*, *adõ (pardon)*, *ai ati (Marie est partie)*, *apẽ (lapin)*, *uj (grenouille)*, *apuš (la bouche)*, *ane (panier)*
lettons: *aiz (maize)*, *aģi (nagi)*, *ajak (vairak)*, *aťiňa (jaciņas)*, enne *(Nenne)*
polonais: *eś (weź!)*, *oć (chodź!)*, *eće (jeszce)*, *oda (woda)*, *au̯a (dała)*
slovènes: *una (luna)*, *aja (Maja)*, *uku (kuku)*, *uda (huda)*
russes: *abika (яблоко)*, *oxkij (лохкий)*, *aoďa (Володя)*, *ampa (лампа)*.

tchèques: *apapani: (nanapaný), eden (jeden), ampa (lampa), emam (nemám), iďim (vidím), opiť (topit), umaš (tu máš), i:žala (žížala), ovať (chovat), i:stek (lístek), ičisťim (vyčistím)*

Comme les initiales exigent une articulation plus ferme que les consonnes intervocaliques, ce phénomène serait-il causé par une sorte d'économie, c'est-à-dire par l'application de la loi du moindre effort, ou le manque d'entraînement de l'articulation? Car on peut supposer que la réalisation de quelques consonnes est plus facile à l'enfant à l'intérieur du mot. Cette idée peut être confirmée par plusieurs de nos constatations. Ainsi le *v* apparaît, d'après nos recherches, qui sont d'ailleurs en plein accord avec les observations de Gvozdev, beaucoup plus tôt à l'intérieur du groupe rythmique qu'à son commencement. Citons ici comme exemple un document tchèque: L'enfant ne réalisait pas (à l'âge 2;0,16) la bilabiale sonore au commencement du mot, en la remplaçant par une autre consonne, tandis qu'il l'articulait sans défaut à l'intérieur du groupe rythmique: *hana (vana)* X *dovani (do vany)*. Dans la langue du même enfant on a pu constater une analogie pour la fixation de la liquide *r*: cette consonne apparaissait tout d'abord à l'intérieur du groupe rythmique après les alvéolaires *(kastrol, trojkolka, drak)*, ensuite dans d'autres combinaisons, et plus tard enfin comme consonne initiale. Cf.: *strkejte* (*r* réalisé correctement) x *la:t* (*rád*; *r* supléé par *l*) — 3;5,26; *la:no (ráno)* × *odra:na (od rána)* — 3;7,11.

Chez un autre enfant tchèque l'articulation correcte du *v* initial s'est fixée six mois après la réalisation correcte de cette consonne à l'intérieur du mot. En ce qui concerne la consonne *r*, elle s'est développée, chez cet enfant, d'une manière analogue à celle que nous avons constatée chez le premier enfant: la consonne vibrante est apparue tout d'abord après les *t* et *d* mais, en dernier lieu, au commencement du mot.

Cependant, le fait que la réalisation des consonnes est plus difficile aux enfants en initiale que dans d'autres positions, peut être confirmé aussi par la pratique des orthophonistes. Ce n'est peut-être pas sans raison que beaucoup d'entre eux réservent l'entraînement aux consonnes initiales pour la dernière étape dans l'acquisition d'une réalisation correcte ou d'une consonne qui s'enregistre nouvellement dans le répertoire phonémique de l'enfant.

Toujours est-il que l'omission de la consonne initiale est un trait caractéristique des débuts du langage des enfants. Cela serait-il une sorte de persistance de la période où l'enfant omettait la consonne qu'il n'avait pas encore acquise? On peut même supposer que les enfants plus âgés remarquent ce phénomène eux-mêmes, aimant savourer, comme l'a souligné Kornej Čukovskij, la force mystérieuse des sons de la parole. Car, en jouant au bébé, dans un style câlin, ils omettent les initiales même à l'époque où ils les réalisent régulièrement dans le langage courant. C'est pourquoi nous avons saisi, dans le style câlin de plusieurs enfants tchèques, les réalisations suivantes: *emam (nemám)*, *ejsek (pejsek)*, *omoc! (pomoc!)*, *usim (musím)*; et, ces exemples pourraient être facilement enrichis par la citation d'autres réalisations semblables.

En concluant, signalons une analogie du fait analysé avec le développement historique de quelques langues où les consonnes initiales se développent d'une autre façon que les consonnes intervocaliques. Ce qui pourrait confirmer notre conviction que l'étude du dévelopement du langage enfantin n'est pas sans importance pour la phonétique et la linguistique générales.

DISCUSSION:

BOREL-MAISONNY: Les consonnes initiales dans le retard de parole sont inégalement atteints et moins que les finales, les *l* et les *r* sont les plus touchées.

Un rencensement serait possible en français à partir des observations des retards de parole. Les résultats seraient variables dans les parlers régionaux.

DEZSÖ: Prof. Ohnesorg has demonstrated a rule by which the phonetic systems of the child language may be obtained from that of adult language. This is an approach different from that accepted in the ontogenetic studies. They try to show the development of the adult system from child language. But the child language has two faces, it is also a language learned from the adults and as such it has its own rules. Now the problem arises: how to confront these two approaches. To answer this question we have to take into consideration: the competence and performance in child language. The rule proposed by prof. Ohnesorg seems to belong to performance. Here it is interesting to point out, that he proposes universal or quasi-universal rules of performance, because no rules of this kind have been proposed earlier.

DVONČOVÁ: Gestatten sie mir eine kurze Bemerkung zu der Initialposition aus dem Slowakischen zu geben. Unsere Kinder haben anstatt *nohavičky* — *havičky* und Anstatt *svetrík* — *etík* gesprochen. Trotzdem sie die Aussprache *n* und *v* gut beherrschten.

PAVLOVIĆ: Les études pédolinguistiques de notre cher Président Ohnesorg ont contribué à bien comprendre la nature du langage de l'enfant. La communication présente nous permet de suivre les interprétations suggestives de la stabilisation des consonnes initiales comparées aux consonnes intervocaliques. L'auteur constate aussi une analogie concernant certaines langues.

RŪĶE-DRAVIŅA: Das Resultat der Untersuchung von Ohnesorg über das Erscheinen von einfachen Konsonanten im Inlaut und Anlaut findet eine Entsprechung in der Bemeisterung der Konsonantengruppen in der intervokalen Stellung und im Anlaut:

Gemäß der Untersuchung von V. Rūķe-Draviņa über lettische und schwedische Kinder (z.B. im Manuskript, mehrmals in Schweden diskutiert) erscheinen die Gruppen -CC- in kindlichen Äußerungen früher als CC-. Die festgestellte Regelmäßigkeit ist ein bedeutungsvoller Beitrag der vergleichenden Pädolinguistik für allgemeine und vergleichende Sprachwissenschaft.

SLAMA-CAZACU: Je trouve que M. Ohnesorg nous a présenté une communication extrêmement importante, car elle ouvre des voies dans la 'linguistique enfantine comparée'. Quelques remarques: (1) on pourrait soutenir l'explication de ce phénomène par la difficulté d'articulation de la consonne en position initiale et non par une perception fautive, en ayant recours aux résultats de la psychologie expérimentale concernant la perception des mots ou la mémoire de courte durée: ceux-ci démontrent que la partie initiale du mot est très importante, qu'elle est retenue assez bien d'habitude. Donc, le phénomène étudié n'est pas dû à un défaut dans la différenciation perceptive, mais dans la réalisation concrète des sons.

(2) Une direction d'études serait celle d'établir un inventaire, en plusieurs langues (et sur des bases statistiques), des consonnes (ou des groupes) en position initiale qui sont omises; si TOUTES les consonnes sont omises, dans toutes les langues, ou seulement CERTAINES consonnes, autres pour chaque langue.

(3) Ce phénomène est plus complexe, car il se combine avec celui du raccourcissement des mots et, il serait intéressant d'étudier ces deux phénomènes en les confrontant l'un à l'autre.

(4) Un problème à étudier en relation avec le fait en question est celui de l'ampleur de la modification structurale du mot. Il y a des mots qui sont peu modifiés (en tchèque: *lampa* > *ampa*), il y en a dont la structure est complètement changée (en français: *grenouille* > *uj*).

(5) On pourrait se poser aussi la question si ces modifications phonétiques

— lorque la structure du mot est complètement modifiée — n'ont pas des implications importantes dans la communication (entre enfant et adulte, ou entre locuteurs enfants). Est-ce que l'enfant emploie certains procédés (linguistiques ou extralinguistiques) et qui seraient différents selon les langues, pour compenser ces déviations qui pourraient empêcher une réalisation normale de la fonction communicative du langage?

(6) On pourrait rapprocher ces faits aussi des modifications phonétiques observées dans l'histoire de chaque langue ou de faits dialectaux, car des phénomènes pareils peuvent être constatés à une certaine étape de l'histoire d'une langue ou dans une région dialectale, parfois périphérique.

(7) Les faits constatés ont une très large portée théorique qui vise en premier lieu ce problème des universaux, ou, entre autres, ils pourraient intéresser aussi la théorie de la COMPÉTENCE/PERFORMANCE (ou dans le système exposé ici par M. Kloster-Jensen de la dichotomie: KINÉMATIQUE/PHONÉMATIQUE), car ils montrent un évident décalage entre ces deux lignes de fonctionnement de la langue.

ZUM ERWERB DER ESTNISCHEN QUANTITÄTSREGELN IM ZWEISPRACHIGEN MILIEU

ELS OKSAAR

(Hamburg)

1. Zu den aktuellen Gebieten, auf denen sich die linguistischen Forschungen der letzten Dezennien zunehmend entfaltet haben, gehört die sprachliche Ontogenese,[1] deren Untersuchung auch von der generativen Transformationsgrammatik gewisse Impulse erhalten hat. Das Interesse für die Erforschung der Sprachkompetenz, der Fähigkeit, die es möglich macht, Sätze zu bilden und zu verstehen, hat zugenommen und gleichzeitig haben es auch die psychologischen und genetischen Fragestellungen zur Problematik des Sprachvermögens.

Schon 1935 hat Karl Bühler betont: "Die einzige Gelegenheit, die wir haben, die menschliche Sprache in statu nascendi zu beobachten, bietet das Kind."[2] Aber erst die Sechziger Jahre haben es verwirklicht, daß die rein linguistisch-theoretischen Forschungen und das Studium des primären Spracherwerbs einander näher gekommen sind. Die Notwendigkeit der Kindersprachenforschung wird von der generativen Transformationsgrammatik u.a. für die Vertiefung der Fragen der universalen Grammatik hervorgehoven.

1.1 Die generative Grammatik hat sich bei der Erforschung der Sprachkompetenz durch den hypothetischen idealen "native speaker" einer völlig homogenen Sprachgemeinschaft zu sehr von der Sprachwirklichkeit isoliert. In der heutigen Lage wird auch die

[1] Zur heutigen Forschungslage s. E. Oksaar, "Zum Spracherwerb des Kindes in zweisprachiger Umgebung", *Folia Linguistica* IV: 3/4, 108-136.

[2] In: *Archiv für die ges. Psychologie* 94, 410.

zweifelsohne sehr wichtige Verbindung zwischen Performanz und Kompetenz vernachlässigt. Ich sehe daher gerade in der wissenschaftlichen Analyse der Kindersprache einen methodischen Weg, die Annäherung der Kompetenz- und Performanzforschung zu erzielen. Für das Studium dynamischer linguistischer Prozesse bietet sie eine breitere empirische Grundlage als die Sprache der Erwachsenen.

2. Wir analysieren im folgenden einige Daten des Erwerbs der estnischen Quantitätsregeln, befassen uns aber auch mit methodischen Fragen. Die Daten sind Tonbandaufnahmen der ersten drei Jahre meines Sohnes Sven in Stockholm entnommen. Die täglichen Sprachen seiner Umgebung waren Estnisch und Schwedisch. Material von fünf Kindern desselben Alters und mit demselben sprachlichen Hintergrund wird als Kontrolle herbeigezogen.

2.1 Als Ausgangspunkt einige Bemerkungen zum Aufbau der phonetischen und phonologischen Einheiten in der Sprache dieses Kindes. Bei seinen ersten Wörtern fällt auf, daß zwei qualitativ verschiedene Konsonanten oder Vokale verwendet werden, jedoch nicht die beiden Verschiedenheiten gleichzeitig. Estn. [mæmma] für [mamma], [ema, æma] 'Mutter', schw. [apa] 'Affe' (7-9 Montae) Diese Feststellung scheint die Beobachtung von Jakobson zu bestätigen: "Das Kind kann im Worte zwei verschiedene Konsonanten oder Vokale verwenden, aber nicht die beiden Differenzen zugleich."[3] Aber schon früh, im 8. Monat ergaben sich Wörter mit anderer Distribution [geba], ein Globalausdruck für Identifikation (Appellfunktion), auch in Verdoppelung: [geba geba]

Die Analyse des Gesamtmaterials legt folgenden psycholinguistischen Schluß nahe und führt uns zur methodischen Vorsicht bei der Analyse der Einzelfragen. Es werden Laute und auch schwierige Lautkombinationen leichter und schneller in den Wörtern gelernt, die das symbolisieren, was das Kind besonders gern hat. Die Automarke *NSU Prinz* wird immer richtig ausgesprochen (30 Monate), sonst *nts* > *ns*; schw. *dillkött* [díltʃøt],

[3] *Kindersprache, Aphasie und allgemeine Lautgesetze* (Uppsala, 1941). Zitiert nach der Edition Suhrkamp Ausgabe (Frankfurt/Main, 1969), S. 120.

seine Lieblingsspeise, ebenso, sonst *tʃ* < *s*. Dies zeigt, daß man auch beim Lauterwerb mit einem Komplex von Steuerungsfaktoren rechnen muß und von Anfang an die Einwirkung des semantischen Systems und des Kommunikativ-Funktionalen bei der Sprache berücksichtigen muß.

2.2 Einen interessanten Fall in der Aneignung der estnischen Phonemstruktur bietet die Analyse der Quantität. In den meisten Sprachen gibt es höchstens zwei phonematisch verschiedene Grade der Quantität. (Sie haben ein binäres System, wodurch z.B. lange und kurze Vokale unterschieden werden.) Im Estnischen begegnen wir nun dem seltenen Fall, daß sowohl Vokale als auch Konsonanten mit drei Quantitätsstufen vorkommen: kurz, lang, extra lang. Die drei Quantitätsstufen haben phonematische Funktion, wie aus folgenden Wörtern hervorgeht. Vokale: 1. Stufe *sada* 'hundert', 2. Stufe: *saada* 'sende' (Imperativ), 3. Stufe: *saada* 'ich will (es) haben'. Konsonanten: 1. Stufe: *lina* 'Laken', 2. Stufe: *linna* 'der Stadt' (Gen.), 3. Stufe: *linna* 'in die Stadt' (Kasus Illativus).

Im Alter von 25-27 Monaten war dieses Quantitätssystem beim Kinde schon voll ausgebildet, während das gesamte Lautsystem zur gleichen Zeit noch nicht ausgebaut war. Das Kind unterschied z.B. die drei Quantitäten des *i* in: (1) *pime* 'dunkel', (2) *piima* 'Milch' (Gen.), (3) *piima* 'Milch' (Part.). Nach dem Alter von 28 Monaten waren keine Abweichungen von der Kompetenz dieses Systems festzustellen. Die Fluktuation vor dieser Zeit betraf die Quantitätsstufe 2 und 3, und kam nie zwischen 1 und den beiden anderen vor.

2.2.1 Diese Feststellungen, die ich bei fünf anderen estnischen Kindern in Stockholm mit ähnlichem Resultat überprüft habe, müssen nun dem bekannten, allgemein akzeptierten Gesetz von Jakobson über den sukzessiven Erwerb des phonematischen Systems einer Sprache in ein anderes Licht rücken. Jakobson stellt fest: "Gegensätze, welche in den Sprachen der Welt verhältnismäßig selten vorkommen, gehören zu den spätesten Erwerbungen des Kindes."[4]

[4] 1941: 76.

Die estnischen Quantitäten sind Gegensätze, welche in den Sprachen der Welt sehr selten vorkommen. Bei den estnischen Kindern gehören sie aber zu den frühesten Erwerbungen. Dies zeigt, daß das Gesetz Jakobsons nicht ausnahmslos ist. Wie ist das zu erklären? Ganz einfach — von der sprachlichen Realität her. Wenn wir von Anfang an bei der Erforschung des Spracherwerbs methodisch so vorgehen, daß die lautlich-phonologische Seite von der semantisch-kommunikativen nicht isoliert betrachtet wird, können wir auch diese Frage beantworten. Die estnischen Quantitätsunterschiede sind dem Kinde von Anfang an wichtig zu erlernen, da sie sehr frequente Wörter (Formen) unterscheiden. Das Kind hört sie auch ständig in seiner Umgebung. Die Frequenzfrage in der Kommunikation hat Jakobson bei seinen Feststellungen nicht berücksichtigt. Wie stark die Quantitätsregeln wirksam sind, geht aus Interferenzfällen hervor. Wenn schwedische Grundformen mit estnischen Endungsmorphemen versehen werden, wird die schwedische Phonemfolge häufig auch dem estnischen Quantitätsgesetz unterworfen: *hälsama* (schw. *hälsa* + estn. *ma* — Inf.-Endung), Stufe 3.

Auch andere Resultate aus neueren Untersuchungen lassen dieses Gesetz (unter psycho- und soziolinguistischen Gesichtspunkten) als überprüfenswert erscheinen. Laut Jakobson — an Benveniste anknüpfend — entwickelt sich entweder *l* oder *r* beim Kinde als einer der letzten Laute, weil es viele Sprachen gibt, die die Distinktion zwischen *l* und *r* nicht kennen.[5] Ruth Weirs Untersuchung zeigt, daß *l* und *r* bei ihrem 2½-jährigen englischsprechenden Sohne schon voll funktionsfähig waren, während es bei anderen Lauten Fluktuationen gab.[6] Eine Statistik des estnischen Materials zeigt, daß in gleichem Alter (30 Monate) *l* dominiert, *r* aber auch voll ausgebildet ist. Rein statistisch ist es so. Aber betrachten wir die Phänomene im Kontext mit anderen Wörtern, so sehen wir, daß die Fluktuationen in vielen Fällen kontextbedingt sind, also assimilatorisch.

[5] 1941: 77.

[6] *Language in the Crib* (= *Janua Linguarum*, ser. maior 14) (The Hague, 1962).

(1) *Pane tuli äla* (richtig: *ära*) 'mach das Licht aus'.
(2) *Onu Berti para kardi ära = onu Bertil parandas kardina ära* 'Onkel Bertil hat die Gardine ausgebessert' (25 Monate).

Im ersten Satz scheint *äla* statt *ära* von dem *l* im vorangehenden *tuli* beeinflußt zu sein, im zweiten erscheint die richtige Form — sehr wahrscheinlich wieder durch die Unterstützung der Distribution der *r*-Laute in der Gesamtäußerung (ansonsten: der Satz weist mehrere reduzierte Varianten der Wörter auf).

3. Zusammenfassend läßt sich feststellen, daß wir bei der Analyse des Erwerbs der phonetischen und phonematischen Systeme die soziokulturellen Merkmale der Umgebung nicht außer Acht lassen dürfen. Dies zeigt uns — und wenn wir das Gesamtmaterial herbeiziehen, sehen wir es noch deutlicher[7] — daß es auch viel zu eng ist, nur von der linguistischen Kompetenz (im Sinne Chomskys) beim Spracherwerb zu sprechen. Der Erwerb von sprachlichen Verhaltensweisen ist mehr als nur die Aneignung der Aussprache, der Lexik und der Grammatik. Er ist gleichzeitig der Erwerb von neuen Verhaltensweisen, die situationsbedingt sind und zur kommunikativen Kompetenz gehören. Erst die kommunikative Kompetenz, die die soziokulturellen und die rein situativen Merkmale berücksichtigt, ermöglicht den Menschen, vorher nie gehörte Sätze zu verstehen.[8] Das Kind erwirbt nicht nur die Fähigkeit, Sätze zu bilden, sondern sie auch zu verstehen. Die Voraussetzungen dazu sind keineswegs dieselben — ein Umstand, den die generative Transformationsgrammatik bei den Erörterungen des Kompetenzbegriffes nicht berücksichtigt hat. Die Methodik der Kindersprachenforschung darf daran nicht vorbeigehen.

[7] Vgl. den Anm. 1 erwähnten Aufsatz.
[8] Siehe hierzu E. Oksaar, "Sprachliche Interferenzen und die kommunikative Kompetenz", erscheint 1972 in *Alf Sommerfelt Gedächtnisschrift* (München, Hueber).

DISCUSSION:

KAPER: Verstehen sie unter 'Kompetenz' etwas anderes als Chomsky unter "competence", wenn sie von 'kommunikativer Kompetenz' sprechen?

KLOSTER-JENSEN: Zur Stärkung von der grundlegenden Idee Oksaars kann angeführt werden, daß sogar eine Abart im Gebrauch von Worttönen in den skandinavischen Sprachen im sehr frühen Alter beherrscht zu werden scheint. Beispiel: *ta på frakken* /ta po fraken/ mit zwei Bedeutungen: 'den Mantel berühen' (Tonem 1), 'den Mantel anziehen' (Tonem 2).

LIŠKA: Have you noticed any interference of intonation patterns between Estonian and Swedish in your child? I presume there must be some.

PAČESOVÁ: (1) At what stage of their speech development do the Estonian-speaking children master the feature of length?

(2) Is there any difference between the emotive and non-emotive speech in children? Do extra-long allophones appear only in emotional expressions?

SIERTSEMA: Underlines the speaker's warning against laying down general laws as to what children say first. Another example is one that refutes the so-called rule that children produce plosives before they produce fricatives. In Dutch one child at least had as its very first word /γof/, the name of a little baby-goat his mother used to call *schoffie* /sxɔfi/ — 'little rogue'.

ANSWERS:

(to Kaper) Ja, denn Chomskys "linguistic competence" gehört zum idealen Sprecher-Hörer in einer völlig homogenen Sprachgemeinschaft. Die Grammatik einer Sprache beschreibt laut ihm diese Kompetenz. Er berücksichtigt aber nicht, dass diese Grammatik höchstens imstande ist, die Sprecherseite zu beschreiben, nicht aber die Hörerseite, also die Merkmale zu erklären, die es ermöglichen, nie zuvor gehörte Sätze zu VERSTEHEN. Die soziokulturellen und situativen Merkmale müssen daher ebenso berücksichtigt werden. Für diesen grösseren Rahmen ist es angebracht, von kommunikativer Kompetenz zu sprechen. Ich verweise auf meinen Aufsatz "Sprachliche Interferenzen und die kommunikative Kompetenz" (Fussnote 7, s. Referat).

(to Liška) The Estonian intonation can dominate in certain cases over the Swedish one; it replaces the Swedish accent II in: *gata, prinsessa.*

(to Pačesová) Whereas the three quantities were mastered at the age of 25-27 months, the mastery of the difference between the two first ones (short-long) was achieved already at the age of 16-18 months. — We can find extra long allophones in all kind of expressions, not only in emotional ones.

THE GROWTH OF PHONEMIC REPERTORY IN CZECH-SPEAKING CHILDREN

JAROSLAVA PAČESOVÁ

(Brno)

The present study is based on the analysis of the speech of children between two and three years from the Brno area in an attempt to find out the degree of completeness of their phonemic system. From the data recorded it is hoped to make assessment of: first, the comparison of the child's phonemic system with that of adults; second, the distribution of phonemes in children in this age group; third, the trend of development of the phonemes in Czech children in comparison with the tendencies observed in the language development in general.

As for VOWELS, the children under observation have all vowels, short and long, which build up the vocalic system in Standard Czech. Not all of them, however, are equally well mastered as far as all distinctive and redundant features are concerned. Only the vowels /a/, /i/, /u/ may be considered as fully mastered phonemes. They are NEUTRAL, based on opposition FRONT VERSUS BACK and HIGH VERSUS LOW. Their phonetic implementation is fairly stable, they are used in their proper positions and are not replaced by other vocalic phonemes, on the contrary, they themselves function as substitutive sounds for other phonemes whose phonetic realization has not yet been perfectly acquired.[1] Besides the stability, the wide distribution of the three vocalic phonemes — and foremost,

[1] Thus, e.g. the phoneme /a/ is used in the place of /o/ and the diphthong /au/, the phoneme /i/ in the place of /e/, the phoneme /u/ in the place of syllabic [r̥] and [l̥].

of the phoneme /a/ — is a typical characteristic of the speech of the observed children. In explaining this we are confronted with the question of whether the vocabulary is selected in order to use these phonemes so frequently or whether the vocabulary so far acquired by this age group accounts for this. When analysing the fundamental stock of words in children in the first stages of their speech development, we see that most of the expressions are — to use Jakobson's terminology — the parental terms,[2] further, the nursery forms and interjections, especially those of onomatopoeic origin. As such they are adapted to the children's phonemic system as well as to their creative abilities. — The situation is rather more complicated with the MID VOWELS /e/ and /o/. Though they do exist in the vocabulary of all the children investigated, their, as yet, imperfectly mastered phonetic realization is reflected in a number of Non-Standard allophones and their mutual overlapping.

The LONG VOWELS represent a less mature stage of development as compared with their short counterparts. Instability shows foremost in length, where the duration of the long vowels does not correspond to the demand in Standard Czech in their being roughly twice as long as the corresponding short vowels. Neither is their quality stabilized. They are rarely neutral, an additional feature of OPEN and CLOSE being attributed especially to the mid vowels /e:/ and /o:/. The implementation of the vocalic chains has not been mastered either. As for DIPHTHONGS,[3] they are monophthongized in most cases and a long vowel is then their substitute. Also in mastery of the HIATUS, children find difficulties. In most cases they avoid it by inserting the hiatic consonant.

To summarize the role of articulatory features in the children's

[2] On this question cf. R. Jakobson's study "Why Mamma and Papa" in *Selected Writings*, I: *Phonological Studies* (The Hague, 1962), pp. 538-45, and the chapter "Mamma and Papa" in O. Jespersen's monograph *Language, its Nature, Development and Origin* (London-New York, 1922), pp. 154-60.

[3] One important fact, however, must be mentioned here in connection with the later acquirement of the diphthongs as compared with the acquirement of the simple vowels in Czech-speaking children: Standard Czech has but one diphthong in the domestic stock of words, i.e. [ou], which, moreover, has not a very high frequency. Naturally, the children in this age group have few posibilities of hearing and little practice in interpreting it.

vowel system, our findings indicate the following: the contrast HIGH VERSUS LOW has been learned well and so have the contrasts FRONT VERSUS BACK and WIDE VERSUS NARROW. The three-fold opposition LOW-MID-HIGH, on the other hand, remains an imperfectly learned feature. Similarly, the feature of ROUNDING shows certain instability both as far as the degree is concerned and in being attributed, beside the back vowels, also to the front /i/ and /e/, whereas only back vowels are rounded in Standard pronunciation. The feature of LENGTH is another distinction the children have not yet acquired well. Instability is shown in the existence of the semi-long, long and extra-long allophones which appear in free variations, while the phonemic value is not attributed to any of them. The children, however, do perceive the length. This shows especially in their replacing the diphthongal chain by means of a long vowel. Were they not able to recognize the feature of length in them, in simplifying them they would, in my opinion, drop one of the vowels analogously as they do in simplifying the consonantal clusters. In summary then, PHONETICALLY, the children's vocalic system, compared to Standard Czech, is far more rich. PHONEMICALLY, on the other hand, it is reduced, as none of the mentioned qualitative or quantitative allophones is used contrastively.

As for CONSONANTS, there is, in accordance with Standard Czech, a 4-term voiced/voiceless system of ORAL STOPS, i.e. /p/-/b/, /t/-/d/, /ť/-/ď/, /k/-/g/, accompanied with three NASALS, i.e. /m/, /n/, /ň/. In terms of features, the STOP ARTICULATION has been learned well. So has the feature of NASALITY. The contrast based on the point of articulation may be considered as but imperfectly-mastered. The front stops exhibit greater stability as compared to the back ones. This is shown in the fact that the velar stops are often replaced by the alveolar ones, thus neutralizing the feature front versus back. It is, however, the feature of VOICING which shows greatest instability.[4] The fluctuation of voiced and voiceless stops takes place at the

4 The loss of the feature of voice and the confusion between voiced and voiceless stops has been noticed by many investigators who have analysed speech development. Among those whose subjects were Czech-speaking children we might mention K. Ohnesorg, cf. *Fonetická studie o dětské řeči*

bilabial, alveolar and palatal points of articulation, word-initially and word-medially. The neutralization of the feature of voice in the final position, obligatory for Standard Czech, is perhaps the explanation why the fluctuation VOICED VERSUS VOICELESS does not occur word-finally. The voiced stops are — as a rule — replaced by their voiceless counterparts.[5] On this basis the predominance of the voiceless stops might be explained. As with vowels, so too with the stops, few additional sound differences, unknown to Standard Czech, appeared in children. ASPIRATION and PALATALIZATION adopted to express the difference between emotive and non-emotive approach might be mentioned here.[6]

Of the existing FRICATIVE PHONEMES in the Czech consonantal

[A Study in Pedophonetics] (Praha, 1948), p. 42; *Idem*, *Druhá fonetická studie o dětské řeči* [Another Study in Pedophonetics] (Brno, 1959), p. 53; *Idem*, "Une contribution à la pédophonétique", *Zeitschrift für Phonetik* 17 (1964), 265-68, further F. Trávníček, *O vývoji dětské řeči* [On Speech Development] (Brno, 1931), p. 150; J. Janko, "Několik postřehů z vývoje dětské mluvy" [A Few Observations in Speech Development], *ČMF* 24 (1938), 131; L. Bartoš, "Observations sur les réalisations phonétiques dans le langage d'un enfant de deux ans", *SPFFBU* A 7 (1959), 6; J. Pačesová, *The Development of Vocabulary in the Child* (Brno, 1968), p. 45. Among Non-Czech authors at least these few should be recalled in this connection: Cl. and W. Stern, *Die Kindersprache* (Leipzig, 1928), p. 334; R. W. and J. B. Albrights, "The Phonology of a Two-year-old Child", *Word* 12:3, (1956), 389; M. Cohen, *Sur l'étude du langage enfantin* (Paris, 1952), pp. 115, 124; J. Feyeux, *L'acquisition du langage et ses retards* (Trévoux, 1932), p. 69; A. Grégoire, "L'apprentissage du langage", *Univ. Liège* LXXIII (1937), 206; M. Grammont, *Observations sur le langage des enfants* (Paris, 1902), p. 69; R. Weir, *Language in the Crib* (The Hague, 1962), p. 47-48, has also examples illustrating the loss of the feature of voicing in her son's speech. Unlike other investigators in this field, she explains it on the basis of environmental influence. She herself, having Czech and German as her native languages, has the neutralization of the feature of voice word-finally in her non-native English as well and suggests that the child imitated her speech in this regard; other examples are, in her opinion, due to the child's imitation of the Danish pronunciation of English which often neutralizes voicing in any position according to the Danish contrastive system.

[5] K. Ohnesorg, cf. *Druhá fonetická studie o dětské řeči* [Another Study in Pedophonetics], p. 23 and J. Janko, "Několik postřehů z vývoje dětské mluvy" [A Few Observations in Speech Development], p. 131 have also the opposite examples where the voiceless consonants are replaced by the voiced ones.

[6] On the question regarding the means of expressing emotional stress cf. V. Machek, *Studie o tvoření výrazů expresivních* [A Study on the Formation of Expressive Formulations] (Praha, 1930), p. 5.

system the phonemes /r/ and /ř/ are absent from the children's system as yet. Except for /j/ and /h/ the learning process of the phonetic implementation of none of the fricatives has been fully accomplished. The number of occurrences of fricatives is, compared to stops, considerably lower. Unlike stops, the fricatives are very often dropped, mostly in consonantal clusters, while in single consonants there is frequent overlapping between the fricative and the corresponding stop-phoneme. In terms of features, the FRICATIVE ARTICULATION is to be considered as an imperfectly-mastered feature. The fluctuation between the stop and the fricative — more frequent in voiced fricatives as compared to the voiceless, reveals the instability. Similarly, the as yet unstable feature of LATERALITY is shown in the overlapping of the phonemes /l/ and /j/. VIBRATIVITY remains a non-mastered feature. The contrast front versus back might be considered as established in fricative consonants. The POINT OF ARTICULATION, however, still shows certain instability. Its misuse gives rise to a number of allophones, untypical of Standard Czech, as bilabial [f], bilabial [w], palatal [l'] and palatalized [s', š', z', ž']. Unstable also is the feature of VOICING. This shows in the fact that the voiced fricatives are often replaced by their voiceless counterparts; also their occurrence is less frequent as compared to other consonants.

As for AFFRICATES, the children have two in their consonantal system, i.e. /c/ and /č/. Neither of them, however, might be considered as well mastered. Instability shows both in MANNER and POINT OF ARTICULATION. The former accounts for the fluctuation between the proper semi-occlusive and fricative consonants, i.e. /c/, /č/ and /s/, /š/ respectively, the latter for the rise of Non-Standard palatalized allophones. The older developmental stage is reflected in some children in their substituting the proper affricates by the corresponding stops /t/ and /ť/.

In summary then, from the six MANNERS OF ARTICULATION upon which the phonemically relevant distinctions of consonantal phonemes in Standard Czech are based, the children have mastered well the OCCLUSIVITY and NASALITY. FRICATIVITY, LATERALITY and SEMI-OCCLUSIVITY, on the other hand, are to be classified as but imperfectly-learned features and no examples which would illustrate the children's mastering the last of the features — the VIBRATIVITY — were recorded. From all consonants which represent the five distinctions, the stops clearly show predominance.

As regards the distinctions based upon the POINTS OF ARTICULATION, the children have mastered, with minor exceptions, all which have phonemic relevance in Standard Czech, i.e. LABIALITY, ALVEOLARITY, PALATALITY and VELO-GLOTTALITY. The manner of articulation then accounts for a further breakdown into the following spheres: BILABIAL and LABIODENTAL, PRAE- and POST-ALVEOLAR, VELAR and LARYNGEAL. The front consonants, i.e. the labials and alveolars, predominate over the back both in stability and distribution. This shows in the replacing of the velar consonants by the corresponding alveolars.[7] The less mature stage of fricatives and affricates is also seen in the less-well-established point of articulation, in contradistinction to stops and nasals.

The PRESENCE OR ABSENCE OF VOICE is another relevant feature in Standard Czech and most of the observed children seem to be aware of it in all consonants. Nevertheless, as in distinctions based on the point of articulation, so too in the distinctions based on voicing, the stops present a more mature stage of speech development compared to fricatives and affricates. In general, the VOICELESS CONSONANTS are more stable and — concomitantly — more frequently used in children.

From what has been said about the vocalic and consonantal

[7] The nasals, however, seem to be an exception. As for the timing, the earlier appearance of alveolar [n] as compared to velar [ŋ] is evident, and data illustrating this fact might be found even in English- and German-speaking children, where both the consonants have the status of independant phonemes. The more easily explicable is the later appearance of [ŋ] in Czech, where this consonant represents but a positional allophone of the phoneme /n/ and where its existence presupposes the realization of consonantal clusters. There are, however, no documents to my knowledge which would illustrate the usage of the alveolar allophone instead of the velar one with a Czech-speaking child at that developmental stage, when he has the velar stops and consonantal clusters in his phonemic repertory. This is rather at odds with R. Jakobson, cf. "Kindersprache, Aphasie und allgemeine Lautgesetze", *Selected Writings*, I: *Phonological Studies* (The Hague, 1962), p. 362, who presupposes the alveolar substitutions for nasals as well.

system in the observed children it follows that the order of phonemic development processes — though not without exceptions, as was suggested in the theses of R. Jakobson — according to the principle of MAXIMUM CONTRAST and starts off by mastering the simple and unmarked features while the differentiated and marked features appear later. The first stage of speech development is initiated by the child's differentiating between a CONSONANT and a VOWEL. This contrast finds its optimal expression when a consonant with a complete lip-closure is opposed to a vowel with a wide frontal opening. As the unmarked features precede the marked ones, the VOICELESS CONSONANT and the SHORT VOWEL are expected in this connection. In particular, the consonant /p/ and the vowel /a/ prove to be the optimal pair. Hence follows the early stability and the wide distribution of the two phonemes in the nursery forms. The principle of the maximum contrast shows also in the successive mastering of the VOCALIC PHONEMES. The fundamental low /a/, stable in all children, forms the first vocalic contrast with the high vowels /i/ and /u/, while the mastering of the vowels in the mid positions comes later. The earlier stability of the short vowels as compared to the long ones, demonstrates the priority of the former group. Similarly, the realization of the SIMPLE VOWEL represents a more mature stage as compared to the realization of the VOWEL CHAIN, regardless of whether the chain is diphthongal or hiatic. — The CONSONANTAL SPHERE is — in Jakobson's observation — broken up into ORAL VERSUS NASAL. Here too, we may find confirmation in our data in the feature of nasality well established in all the investigated children as well as in the wide distribution of the nasals in the first words of the child. Fairly stable in our data also, is the contrast LABIAL VERSUS ALVEOLAR which again Jakobson ranks as the next established in the consonantal phonemes in general. The contrast FRONT VERSUS BACK, or in particular, ALVEOLAR VERSUS PALATAL and ALVEOLAR VERSUS VELAR evidently belongs to the later stages of language acquisition. The lesser stability of palatal and velar articulation, the mutual overlapping of the consonants /t/-/k/ and /t/-/ť/ and high preponderance of the phonemes in the front of the mouth cavity provide clear evidence of the priority of the front consonants as compared to the back ones. As regards the contrasts based on the manner of articulation, the STOP PHONEMES clearly predominate both with regard to their phonetic implementation and their distribution and stability in the first words of the child as a single consonant and in consonantal clusters. The FRICATIVES, on the other hand, represent the less mature stage and appear, and are gradually stabilized, in the points of articulation where the corresponding stops are already firmly established. The lesser stability of a fricative as compared to a stop, shows moreover in the consonantal clusters. It is almost exclusively the stop, which is preserved in simplifying while the fricative is dropped. Similarly, the AFFRICATES — as the representatives of the semi-occlusive articulation — appear and are stabilized only after the corresponding fricatives are well learned. In the earliest stage they are replaced by stops, later on by fricatives before they reach the proper semi-occlusive implementation. The succession of phonological acquisitions, stops — fricatives — affricates, is clearly manifested in our data. The non-existence of the VIBRANTS in children between two and three is in agreement with Jakobson's statement that the distinctive features comparatively rare in the languages of the world, are among the latest phonological acquisitions of the child. The priority of unmarked features is manifested in the earlier stability and wider distribution of the voiceless consonants in the child's vocabulary as it is in languages in general.[8]

Of the vast problematics which child language offers, we have summarized here what we have found most characteristic in the

[8] There are, nevertheless, some exceptions to this rule, e.g. the preponderance of the voiced /h/ over the voiceless /x/ in Standard Czech which shows also in Czech-speaking children's inventory. This fact, however, is to be ascribed to the special and shorter history of this contrastive pair in the development of the language system in Czech, as has already rightly been pointed out by N. S. Trubetzkoy in his *Grundzüge der Phonologie*, cf. *TCLP* 7 (Praha, 1939). — Another exception concerns the pair /f/ — /v/. The peripheral character of the voiceless /f/ — this phoneme appears only in loan-words and interjections of onomatopoeic origin or in words derived from such interjections, cf. *J. Vachek*", On Peripheral Phonemes of Modern English", *Brno Studies in English* 4 (1964), p. 8 — and the wide distribution of the voiced /v/ in the common Czech word stock accounts for this discrepancy.

children investigated. Many observations found in the relevant literature in this field are confirmed in our findings, some are not. It is hoped, however, that the present article brings further evidence of the fact that the speech of children and above all, their approach to the mother tongue in mastering it, progresses in accordance with the tendencies observed in the languages of the world in general.

DISCUSSION:

KACZMAREK: Die These Jakobsons über die Entwicklung der Phoneme in der Kindersprache, die Jakobson festgelegt hat, hat Frau Pačesová in ihrem Vortrage bestimmt erhärtet. Aber man muß im Auge haben, daß es, wie in der polnischen Sprache J. Kania festgestellt hat, einige Abweichungen gibt. Diese, und vielleicht noch einige andere, die in anderen Sprachen vorkommen können, müssen, wenn wir ein allgemeines Gesetz über die Entwicklung der Phoneme in der Kindersprache formulieren wollen, in Betracht genommen werden.

ON THE PSYCHOLINGUISTIC STUDY OF LANGUAGE ACQUISITION

JAN PRŮCHA

(Praha)

The study of child language acquisition has a long tradition dating as far back as last century. In the course of these studies, considerable amount of material and empirical data has been compiled, concerning development of children's speech in various types of languages, as well as the characteristics of children's speech from the viewpoint of various structural levels (particularly phonetic and lexical) of one and the same language. Research of children's speech in all its recorded history has not been confined to purely linguistic aspects — on the contrary, apart from linguists — especially phoneticians, children's speech has been studied in the first place by psychologists and psychiatrists, pedagogues, logopedians etc.[1] In spite of these tremendous efforts exerted by a number of researchers, "in spite of the whole wealth of works on the development of speech which came out in the course of the past decades, we have not yet in fact arrived to what constitutes the core of the problem, and particularly to the explanation of the essence of speech development" (Shugar, 1969). We do not intend to analyse the causes of this state — it is clear that they lie first of all in the complexity and extent of the given phenomenon, for whose analysis there are not yet sufficiently adequate means of

[1] An extensive survey and bibliography of European works on child language acquisition in the years 1920-1968 was published by Slama-Cazacu (1969). For survey of American works in last decade see Jenkins (1969) and Shugar (1969). Slobin (1966) worked out a survey of Soviet studies on child language.

scientific investigation. This paper is meant to briefly evaluate a new approach to the study of language acquisition which has been finding application in one branch of the present psycholinguistic theory, and on the basis of criticism of this approach and ensuing hypotheses, to outline other, more complex conception of the study of language acquisition in children.

It is generally known that the existing psycholinguistics regards as its aim the explanation of two basic processes — i.e. the use of language, and language acquisition by an individual person. During the past decade, psycholinguistic theory brought out very intensive tendencies basing the explanation of these two processes on linguistic models created by N. Chomsky and generativist theorists. A specific trend has thus gradually developed in psycholinguistics, being sometimes called 'transformational psycholinguistics'. Besides its main concern, i.e. so called linguistic competence of adult language users, it studies also the linguistic competence of children and its development. In Chomsky's original conception (cf. e.g. Chomsky, 1964), the language acquisition model was conceived as a device, its input being represented by primary linguistic data, and its output by the grammar of the language. This means that a child acquiring a language in a certain sense 'creates' the grammar of that language which surrounds him. His linguistic competence (i.e. the knowledge of the structure and rules of the language) gradually develops, and approaches the linguistic competence of adults.

This conception of Chomsky was later developed in a number of theoretical and even experimental works, especially by American psychologists and psycholinguists. The main features of the theory of language acquisition, formed within the framework of transformational psycholinguistics, are as follows.[2]

(1) The assumption is that the linguistic competence of a child develops on the basis of 'a biologically-founded, inborn capacity for language acquisition' (see Lenneberg, 1967; McNeil, 1966; 1968).

2 The 'transformational approach' to the study of child language acquisition is represented by such works as Smith and Miller (1966), McNeil (1966; 1968), Klima and Bellugi (1966), Ervin (1966), C. Chomsky (1969), Menyuk (1969).

While the behaviourist theory interprets language acquisition as a formation of language habits during the process of learning, transformational psycholinguistics denies the influence of learning on language acquisition. It presents as one of the proofs of this hypothesis, the fact that a child is capable of understanding and producing such sentences and grammatical structures which he has never come across with before. Ervin (1966) for instance states that children can create grammatical forms (distinguishing e.g. number, tense etc.) of words which they do not yet know, using the analogy of linguistic models. This fact is interpreted as an application by the child of innate structures and rules to linguistic material which he encounters.

(2) Transformational psycholinguistics also attempts to explain the nature of this ability of man for language acquisition. Theie are views that the inborn capacity assumes the form of linguistic universals, i.e. is common to all types of human languages. In the transformational psycholinguistic conception, the study of linguistic universals is identified with the study of the features of generative grammar. It is therefore assumed that the inborn linguistic capacity includes as one of its constituents 'basis grammatical relations' (i.e. for instance the relationship between subject and predicate, between predicate and object, etc.). A theoretical substantiation of this character of linguistic competence, maintained mainly by McNeil (1966), leans also on experimental findings concerning the ways of the grammar acquisition of various languages in children (see esp. Slobin, 1969; 1970).

According to the theory of transformational psycholinguistics, linguistic capacity is part of a general cognitive capacity of man which is also innate. Lenneberg (1967) provides arguments according to which linguistic capacity is conditioned by biological and anatomical-physiological factors. However, he rejects a hypothesis saying that in its essence man's ability to speak is dependent on such general factors of ontogenetic development as the growth of individual's intelligence, the increase of brain-weight etc. It is more likely that we have something to do with so far unknown capacities specific to the human kind. In the same way

Vereščagin (1969) advocates a thesis on the independent nature of linguistic ability.

(3) It is assumed that language acquisition is a relatively rapid process beginning at the age of about one and ending approximately at the age of three or four. "In fact, it may be that strictly linguistic acquisition is completed by age three or so. Further development may reflect lifting of performance restrictions and general cognitive growth, without adding anything basically new to the fundamental structures of syntactic competence" (Slobin, 1969; 11-12). This 'astonishing speed' of language acquisition is accounted for by the very fact that a child starts by acquiring those linguistic phenomena which are universal (i.e. those which determine the general structure of human languages independent of various physical, cultural and other circumstances). The direction of special investigations then follows from these pre-assumptions: they mostly focus in the acquisition of syntax in the early stages of child speech development.

Even though the theory of language acquisition elaborated by transformational psycholinguistics on the basis of Chomsky's ideas has become very popular, and is being further developed in a number of works, it is necessary to point out that it represents a mere hypothesis whose validity has not been verified, and which reveals serious shortcomings. Let us consider those critical objections against this theory which have essential significance for an adequate study of language acquisition.[3]

Let us first of all deal with the nature of the process in which a child forms his linguistic competence. Behaviourist psychology of language and behaviouristically oriented psycholinguistics conceive language acquisition as a formation of a system of speech habits based on external reinforcement, i.e. essentially according to the S-R model. The whole process of language acquisition is thus interpreted merely on the basis on mechanisms of learning and associative operations. Serious objections agains this behaviourist

[3] We shall leave aside basic problems concerning the conception of so-called linguistic competence and the relationship between models of linguistic competence and models of linguistic performance. We have discussed such problems in another work (Průcha, 1970).

conception were published in an extensive review by N. Chomsky (1959) on Skinner's book *Verbal Behavior* (1957). Chomsky's critical views then became the theoretical credo of transformational psycholinguists. Their viewpoint is most explicitly expressed in contributions of Fodor, Garrett, McNeil and others in *Verbal Behavior and General Behavior Theory* (Dixon and Horton, 1968): Any psychological theory explaining behaviour on the basis of S-R constructs and "the doctrinal theory of associanism" is principally inadequate to the model of language acquisition.[4]

This absolute rejection of the influence of learning on the acquisition of linguistic competence, and attempts to explain language acquisition merely by innate capacities, are extremities which can hardly hold up. What is lacking first of all are direct proofs of the pre-assumption that various mechanisms of man's linguistic ability represent purely genetic products. Transformational psycholinguists' hypothesis regarding man's linguistic capacity as innate as a whole, is also disputable. This 'innateness' may apply only to certain constituents of linguistic capacity. The question then is whether it is right to assume the immunity of the linguistic ability from various external factors influencing man's development. It is on the contrary possible to assume the liability of the linguistic ability to such factors, and a functional connection between speech mechanisms and other mechanisms constituting behaviour. Leont'jev (1969) points out that "provided a child possesses a certain capacity unaccountable for on the basis of our knowledge of the genesis and mechanisms of his speech activity, it does not automatically follow that we recognize this capacity is given a priori" (p. 130).

There are many outstanding psychologists (Osgood, Carroll, Jenkins, etc.) who explain language acquisition on the basis of

4 For the sake of preciseness it is necessary to say that there exists a number of variants of the psychological theory of learning, whose interpretations of language behaviour differ in certain points: starting with the single-stage S-R theory (Pavlov, Skinner) and proceeding to the mediation theories (Osgood, Jenkins etc.) which are quite successful particularly in the explanation of semantic aspects of language acquisition. These differences are completely ignored by critics of behaviourism (cf. Osgood, 1968).

variously modified theories of learning. There is no doubt that these theories have considerable explicative strength, e.g. concerning the acquisition of the meaning of various language symbols. Numerous tentative proofs have been published confirming that the acquisition of the meaning of symbols can be explained on the basis of simple principles of conditioning (see eg. Staats, 1968). On the other hand, however, learning thories do not provide a sufficient explanation of a child's acquiring the complex of grammatical rules.

One of the essential shortcomings of the transformational model of language acquisition consists in its being confined only to the grammatical (syntactic) component, rather neglecting the semantic component, although recognizing its significance. The reasons are apparently formal: the linguistic theory of transformational grammar has been worked out most thoroughly in the field of syntax, while attempts to apply it to the semantics of the word and sentence have been less successful. Another shortcoming of the transformational model concerns the description of linguistic competence it provides, which is incomplete, i.e. does not actually reflect the allround linguistic competence of a child. It is generally known that a child in his speech development is earlier and to a higher degree capable of understanding than of expressing utterances. The existing description of linguistic competence based on a certain set of spontaneous or elicited utterances of a child therefore take into account merely the child's speech, but not his capacity of comprehension. It then follows that it is impossible to make any affirmations on the state of linguistic competence in its completeness exclusively on the basis of active speech, and that it is necessary to work out methods of the study of a child's comprehension processes.

What continues to be disputed is the presumable 'astonishing speed' in which language acquisition progresses. Certain recently published works show that the development of a child's linguistic competence is not terminated at the age of three or four, as a number of transformational psycholinguists believe, moreover, not even its syntactic component. It was particularly the research work of Menyuk (1969) on the syntax of 2-7 year-old children, and of C. Chomsky (1969) dealing with the acquisition of syntactic structures in 5-10-year-old children, which clearly proved that the development of linguistic competence does not end in early childhood, but continues long during the school age. "Contrary to the commonly held view that a child has mastered the structures of his native language by the time he reaches the age of 6,[5] we find that active syntactic acquisition is taking place up to the age of 9 and perhaps even beyond" (C. Chomsky, 1969; 121).

Very interesting critical objections against the transformational model of language acquisition have been brought up by Soviet researcher Ter-Minasova (1968). Using material of Russian, gathered among Moscow's children population, she shows that "logical" principle of the "creation of language" according to "inborn" models becomes an impediment in the child's speech development: forms generated according to productive models contradict real facts of speech. "A child at every step encounters innumerable exceptions — semi-productive or even completely unproductive models ... The adults must permanently correct him, breaking arranged logical structures which cannot hold up in the vast number of encounters with the reality of speech communication." The child gradually copes with this problem by means of experimenting and making mistakes, until he fully acquires also those 'illogical' forms of the adults' speech.

Concerning the whole controversy between 'behaviourist psycholinguistics' and 'transformational psycholinguistics', we find adequate the following conclusion: Nowadays it is not possible to declare either of the theories better than the other for the study of language acquisition, for reasons ensuing from their essentially different directions. While transformational psycholinguistics attempts the explanation of the acquisition of linguistic competence (in the sense of the speaker's language knowledge), eliminating all characteristics of the functioning of this competence, behaviourist psycholinguistics studies a real language behaviour, i.e. the linguistic per-

[5] Many transformational psycholinguists suppose a child finishes acquiring of syntactic structures he reaches even the age 3 or 4. (Cf. e.g. McNeil, 1966; Slobin, 1969).

formance affected by various physical, contextual and other restrictions. However, an absolute rejection of one of the theories (e.g. behaviourist, actually rejected by 'transformationalists'), and a monopolization of the other, little contributes to the advancement of science.

However, there arises a question whether a theory based merely on the ascertainment of linguistic competence, can provide sufficient prospects for the study of speech development. The ultimate aim of developmental psycholinguistics is an adequate model of the development of language behaviour. Today it is quite obvious that man's language behaviour is regulated not only by purely linguistic and individual psychological factors, but also by extralinguistic and social psychological ones. Language acquisition represents a communication process, and in its essense language behaviour includes wider than merely linguistic competence, namely communicative competence. This communicative competence lies in the knowledge of rules concerning the use of various forms of language, as well as of various non-verbal codes, in different social situations. Linguistic competence appears to be only one of the constituents of this more general communicative competence. In our opinion, the study of language acquisition cannot be divorced from all contextual (in a wide sense), social and socio-cultural conditions under which the given process takes place. This 'communicative conception' which has been asserted by certain contemporary psycholinguists (e.g. Rommetveit, 1968; Leont'jev, 1969; Slama-Cazacu, 1968, 1970) is after all self-evident — more or less — in earlier as well as more recent descriptive works on children's speech. The influence of the contextual factor (situation) on children's speech has been clearly recognized. Ohnesorg (1959) for instance provides a classification of so called phonic styles in children's speech. He found that the features of child utterances are often "in a direct dependence on the situation in which the utterance took place, and on the child's personal attitude to his own utterance" (Ohnesorg, 1959; 59).

More recent experimental findings of Popova (1968) show that as early as the age of eighteen months, the characteristics of a child's speech activity depend on the types of interaction situations in which the child's communication with adults takes place. A very detailed treatment to the 'communicative conception' of child language acquisition was given by Slama-Cazacu (1961) in her monograph *Dialogul la copii.* It has been proved that the main function of children's utterances is communicative. The need of communication appears even before the child acquires language. That is why elements of dialogue appear in the earliest period of his speech development. Slama-Cazacu's extensive experimental research brings a proof of the fact that dialogue is from the ontogenetic point of view older than monologue, being recorded in children's speech as early as the age of 2-3. This finding is connected with the problem of so called egocentrism of children's speech. Slama-Cazacu's results have fully confirmed findings made by Vygotsky (1934; 1962) and others, denying the view that egocentrism is characteristic for children's speech (which means a limitation of the communicative function), once maintained by Piaget and others.

Dialogue represents the actual (and in fact prevailing) form of a child's linguistic behaviour, and the study of language acquisition cannot be therefore divorced from all relevant conditions in which the speech acts occur. This calls for the study of a whole number of problems connected with a child's communicative behaviour. Let us mention examples of such problems:[6]

(1) The contents of a child utterances, and types of communication contexts of these contents.

(2) The relationship between particular types of communication contexts, and the creation of a child's role-taking skills as well as general socialization.

(3) The relationship between the social structure of a particular community, its norms, attitudes etc., and the acquisition of various language styles.

[6] The study of these problems is beyond the possibilities of developmental psycholinguistics as it is generally understood, and requires such conception as that finding application in the theory of the social psychology of language (or — in terms of Slama-Cazacu, 1970 — 'psycho-socio-linguistics'), which is only just being created (see esp. Moscovici, 1967).

(4) The characteristics of interaction and communication processes between parents and children.

(5) The development of children's speech in dependence on the conditions of their education (family, collective and variously combined), and on sociocultural and other characteristics of educational environment.

(6) The 'communication profile' of child, i.e. the intensity and frequency of his communicative contacts with other children, with parents and relatives, other adults etc.

(7) The development of a child's speech under the influence of various mass media, school education etc.

Certain of these problems have already started attracting the attention of educational and social psychologists, sociologists of mass communication, and others. Number of problems, however, remain intact, particularly their linguistic aspects, although a consistent linguistic analysis of child's language behaviour constitutes one of the pre-requisites of an adequate theory of language acquisition. There is a need of a deeper inquiry into such phenomena as, for instance, chronological regularities in the acquisition of the elements on particular levels of a language system (cf. certain suggestions in Slobin, 1970). Another linguistic problem is the elaboration of methods serving objective measuring of the level of language acquisition reached by a child. It is evident that some of the methods which have been so far used (quantitative indices as well as tests) are not objective (cf. Lawton, 1968).

I hope it is quite clear that the study of above mentioned problems requires developmental psycholinguistics to be more connected with social psychology and sociology than it is made till now. The relationship between language acquisition and social backgrounds was after all recognized and empirically confirmed as early as the twenties and thirties, and was recently submitted to tentative investigation by Lawton (1968) and others. It appears, for instance, that numerous differences recorded in speech of children of different age, social backgrounds etc., do not consist merely in linguistic competence, but in communicative competence. It also appears that communicative competence continues developing also in late childhood and even in adult age. This necessitates the formation of a theory of continual development of man's communicative behaviour.

BIBLIOGRAPHY

Chomsky, C.,
1969 *The Acquisition of Syntax in Children from* 5 *to* 10 (Cambridge, Mass., The M.I.T. Press).

Chomsky, N.,
1959 A review of B. F. Skinner's *Verbal Behavior*, *Language* 35, 26-58.
1964 *Current Issues in Linguistic Theory* (The Hague, Mouton).

Dixon, T. R., and D. L. Horton, (eds.)
1968 *Verbal Behavior and General Behavior Theory* (Englewood Cliffs, N. J., Prentice-Hall).

Ervin, S. M.,
1966 "*Imitation and Structural Change in Children's Language*", *New Directions in the Study of Language*, ed. by E. H. Lenneberg (Cambridge, Mass.), 163-190.

Jenkins, J. J.,
1969 "The Acquisition of Language", *Handbook of Socialization Theory and Research*, ed. by D. A. Goslin (Chicago), 661-86.

Klima, E. S., and U. Bellugi,
1966 "Syntactic Regularities in the Speech of Children", *Psycholinguistics Papers*, ed. by J. Lyons and R. J. Wales (Edinburgh), 183-206.

Lawton, D.,
1968 *Social Class, Language, and Education* (London).

Lenneberg, E. H.,
1967 *Biological Foundations of Language* (New York).

Leont'jev, A. A.,
1969 *Psixolingvističeskije jedinicy i poroždenije rečevogo vyskazyvanija* (Moskva).

McNeil, D.,
1966 "The Creation of Language by Children", *Psycholinguistics Papers*, ed. by J. Lyons and R. J. Wales (Edinburgh), 99-114.
1968 "The Creation of Language", *Language*, ed. by R. C. Oldfield and J. C. Marshall (London), 21-30.

Menyuk, P.,
1969 *Sentences Children Use* (Cambridge, Mass., The M.I.T. Press).

Moscovici, S.,
1967 "*Communication Processes and the Properties of Language*", *Advances in Experimental Social Psychology*, 3, ed. by L. Berkowitz (New York-London), 225-270.

Ohnesorg, K.,
1959 *Druhá fonetická studie o dětské řeči* (Brno).

Osgood, Ch. E.,
1968 "Toward Wedding of Insufficiencies", in: Dixon, and Horton, 1968: 495-519.

Popova, M. I.,
1968 "Osobennosti rečevyx projavlenij detej pervogo polugodija vtorogo goda žizni", *Voprosy psixologii* 14, 116-22.
Průcha, J.,
1970 "Problems of Generative Model in Psycholinguistics", *Recueil linguistique de Bratislava* IV (in press).
Rommetveit, R.,
1968 *Words, Meanings, and Messages* (New York-London-Oslo).
Shugar, G.,
1969 "Nowe kierunki badań amerykańskich nad rozwojem mowy dziecka", *Psychologia wychowawcza* XII:2, 166-188.
Skinner, B. F.,
1957 *Verbal Behavior* (New York).
Slama-Cazacu, T.,
1961 *Dialogul la copii* (Bucureşti).
1968 *Introducere in psiholingvistică* (Bucureşti).
1969 "Studiile Europene asupra limbajului copilului (1920-1968)", part I-IV, *Studii şi cercetari lingvistice* XX: 2, 143-50; 3, 247-74; 4, 361-72; 5, 479-508.
1970 "*Code Levels, Interdisciplinary Approach, and the Object of Psycholinguistics*", *Rev.roum.sci.sociales-psychol.* 14:1, 87-96.
Slobin, D.,
1966 "Soviet Methods of Investigating Child Language", *The Genesis of Language*, ed. by F. Smith and G. A. Miller (Cambridge, Mass.), 361-386.
1969 *Universals of Grammatical Development in Children* (= *Working Paper* 22, Language-Beh. Res. Lab., Univ. of Calif.) (Berkeley).
1970 *Suggested Universals in the Ontogenesis of Grammar* (= *Working Paper* 32, Language-Beh. Res. Lab., Univ. of Calif.) (Berkeley).
Smith, F., and G. A. Miller, (eds.)
1966 *The Genesis of Language: A Psycholinguistic Approach* (Cambridge, Mass.).
Staats, A. W.,
1968 *Learning, Language and Cognition* (New York).
Ter-Minasova, S. G.,
1968 "K probleme detskoj reči", *Voprosy psixologii* 14, 94-101.
Vereščagin, E. M.,
1969 *Psixologičeskaja i metodičeskaja xarakteristika dvujazyčija* (Moskva)
Vygotsky, L. S.,
1934 *Myšlenije i reč'* (Moskva). *Thought and Language*, (Cambridge, Mass., 1962).

DISCUSSION:

SHUGAR: It is difficult to evaluate American psycholinguistic trends without a report on the findings produced. A syntactic structure of children's two-word utterances has been discovered with the basis of English language studies. Does this hold true for other languages? We do not know. It is also difficult to evaluate this trend because it is moving so rapidly. Already syntactic research alone in language acquisition has been replaced by a syntactic-semantic approach (cf. Slobin, 1970; Bronowski and Bellugi, 1970) which is essentially a cognitive approach to language study. Further, while innateness theory is widely hypothesized, there is no denial of the learning function. Since what is learned is now regarded as rules governing linguistic performance (i.e. competence), the central question remains as: how is competence acquired, how much via innate structures, how much via rule-learning and in what interaction. Some American psycholinguists maintain the broader framework of communicative competence as an orienting program (cf. Slobin's *Field Manual for Study of Child Language*).

ANSWER: (1) Investigations based on the transformational model of language acquisition are mostly concentrated with the study of syntax acquisition — they do not deal with semantics (see P. Menyuk, 1969; C. Chomsky, 1969).

(2) Rejecting all learning theories and behaviorism made by 'transformational psycholinguists' does not support the development of scientific knowledge in the field of language acquisition.

SLAMA-CAZACU: There are some fascinating topics of present-day linguistics that are discussed in this paper (such as 'competence' or the 'innateness of language', etc.), Miss Shugar offering personal opinion on them. Concerning her remarks, I would say that:

(1) What is specific to transformational linguistics is conferred to verbal utterance, and that hence, in speaking a wider concept of 'communicative competence' (that would also include nonverbal cues, etc.) would be contradictory to the basic principles of generative-transformational analysis.

(2) In fact transformationalists who sustain a nativistic view do not generally even use the word 'learning' (which is usually replaced by the term 'acquisition'); anyhow, stress is put upon demonstrating that innate aspects do exist and are important, more than upon speaking about 'learning' (or in showing what still remains to be considered as being 'learned');

(3) Concentrating upon one single topic or area present advantages from the scientific methodological point of view, but also some 'dangerous' shortcomings — as the field of investigation is not only limited, but it also might be demonstrated by ignoring some important determinant factors with reference in other areas.

THE EMERGENCE OF AFFIRMATION AND NEGATION IN CHILD LANGUAGE

SOME UNIVERSAL AND LANGUAGE-RESTRICTED CHARACTERISTICS

VELTA RŪĶE-DRAVIŅA

(Stockholm)

1. Negation and affirmation can be expressed by (a) gestures, (b) gestures in combination with spoken words, or (c) articulate speech only. Very often all these types are used side by side in the same language, the actual choice of type depending on the concrete speech situation.

Both gesture speech and the articulated forms are arbitrary and vary from language to language, i.e. different gestures serve as expression for negation, and the gesture which means 'yes' in one language may be used as a negative sign in some other language.[1] Even SPOKEN negative and affirmative utterances vary widely, cf. the use of simple or multiple negation as well as differences between negation models used in German, Russian, English etc., as:

[1] The most common sign for 'yes' is a nod, but the Bulgarian, Greek and Turkish 'yes' is expressed by rolling the head from side to side, and by an upward jerk of the head among Abyssinian speakers. The common gesture for 'no' is a shake of the head, but there are many exceptions: the Greek negation by an upward jerk of the head means the opposite ('yes') in Abyssinian. Cf. Theodore Brun, *The International Dictionary of Sign Language* (London, 1969).

(a) Germ. *sie schrieben nicht*, lit. 'they wrote not'

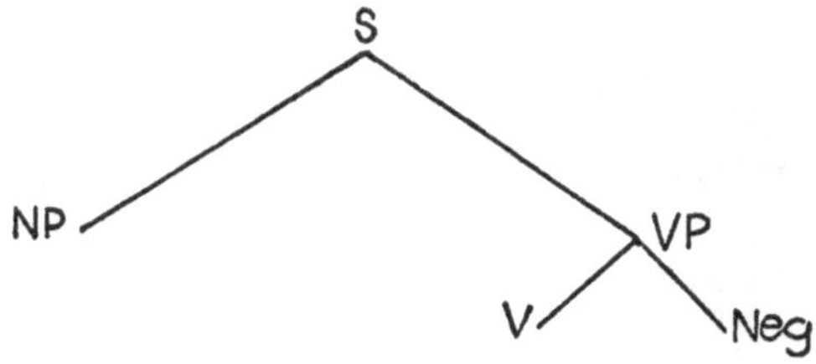

(b) Russ. *oni ne pisali*, lit. 'they not wrote'

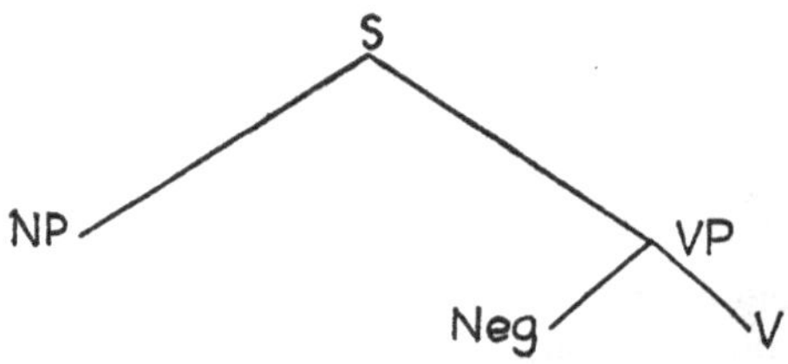

(c) Eng. *they did not write.*

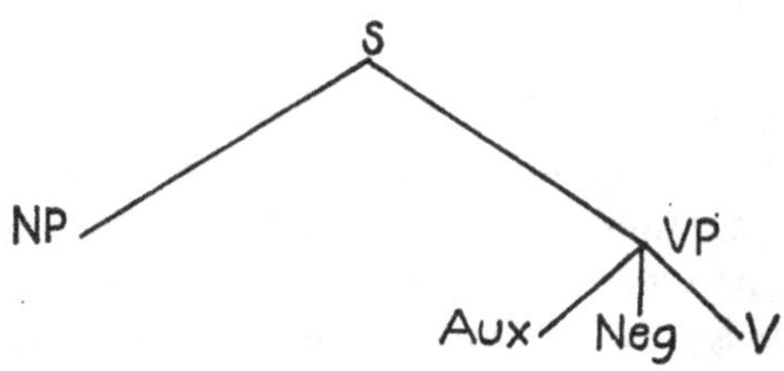

Figure 1.

The main problems in this connection from the general linguistic and pedolinguistic point of view are these:

(1) Do young children learn to express negation through gestures earlier than they begin to use the articulate negative utterances, or at the same time?

(2) Are the negative systems used in different languages equally intricate, i.e. do Russian, English, Latvian, etc. children spend the same time on acquisition of the negative utterances in their native languages?

(3) At what stage of their mental development does the opposition of negation and affirmation appear in the active speech of young children? Which other linguistic phenomena are usually learned by the young child at this time?

2. The only special study limited to the use of negation forms in English children is that of U. Bellugi, "The Emergence of Inflections and Negative Systems in the Speech of Two Children", *Papers presented at New England Psychol. Assn.* (1964). Isolated references to the emergence of negation systems are found in publications about speech development in young children with Russian, Czech, German, English-German, Latvian etc. as native tongues (Gvozdev, Ohnesorg, Stern, Leopold, Rūķe-Draviņa). In recent years, a lively interest in negative and affirmative linguistic systems has arisen also among psychologists (e.g. George A. Miller) through investigation of the problem whether negative utterances are easier or more difficult to grasp and to produce than affirmative ones.

My paper consists of two parts:

In part I previously unpublished material is presented concerning the emergence of affirmation and negation in two Latvian children (a boy and a girl).

In part II this Latvian material is compared with facts published in investigations by pedolinguists concerning other languages, in order to answer the above-mentioned questions and to find out

(a) which phenomena in the speech development process are common to all children irrespective of their native language, and

(b) which phenomena are language-specific.

3. The negation system in Standard Latvian (which was the language used in the environment of the two Latvian children observed) is — in general outline — similar to Czech and Russian, only somewhat simpler.

[nē] or — in colloquial Latvian — [ḿ-m] corresponds to the English 'no'; the corresponding gesture is a shake of the head. The affirmative interjection is [jā] or — in colloquial speech — [m-ḿ]; the corresponding gesture is a nod.

In answers to 'yes'-'no' questions the affirmative and negative [jā] and [nē] can be combined with or replaced by the essential word in this connection (in negative or affirmative form), e.g. *Vai Marija mājā?* 'Is Mary at home?' — *Mājā* 'At home', or *Jā* 'Yes', or *Jā, (ir) mājā* 'Yes, (is) at home'; and — with the negation: *Nav mājā* 'Is not at home', or *Nē* 'No', or *Nē, nav mājā* 'No, is not at home'.

Latvian has only one negative particle [ne-], used with verbs, nouns, adjectives, pronouns and adverbs: *nevaru* 'I cannot', *nelaime* 'bad luck, misfortune', *nelaimīgs* 'unlucky', *nekas* [ne·kas] 'nothing', *nekad* [ne·kad] 'never'.

Ne- forms a single unit with the verb, noun and adjective, and all these negative forms have the stress on the first syllable of the word; but *ne-* in negative pronouns and adverbs is unstressed. In compound tense forms (perfect etc.) the negation *ne-* stands before the auxiliary verb, not before the participle: *es esmu lasījis* 'I have read': *es neesmu lasījis* 'I not-have read' (= 'I have not read').

The negative opposite to *ir* 'is' is *nav* 'is not', but — as Latvian has only one verb for 'to be' and 'to have' — this *nav* in some syntactic constructions may mean 'has not' as well.

In antithesis the verb need not be repeated a second time but may be replaced by the negation *ne* or (dial.) *nē*, cf. *tu vari, es - ne* 'you can, I [can]not'.

Characteristic of Latvian is multiple negation, so that in a negative sentence every part of speech to be negated must have a negative particle: *neviens nekā nezina* [*ne·viens ne·kā nezina*] 'nobody nothing not-knows' (= 'nobody knows anything'). The negative conjunction is *ne - ne* or *nedz - nedz* 'neither - nor'. In the answer to a negative question the word [nē] 'no' may correspond to the English 'yes': *tu taču neesi skolotājs? — nē, es esmu!* 'Are you not a teacher?' (lit. 'no, I am') (= 'yes, I am [a teacher]').

4. An outline of the emergence of negative and affirmative utterances in the children observed is presented in Tables I and II, and

is based on a longitudinal study of a Latvian boy and girl. The basic material consisted of daily records of all significant phenomena in the process of speech development from 0;1 until 4;0 and 0;1 until 6;0 respectively (and later, in more general terms) and frequent tape recordings; in addition, a general report about the linguistic stage reached by the child was written every month. Earlier publications by the author about the speech development of these Latvian children, including a large number of negative utterances, are (as concerns the emergence of declension and conjugation systems): "Zur Entstehung der Flexion in der Kindersprache", *IJSlLP* 1/2 (1959), 203-222 and (concerning syntax) *Zur Sprachentwicklung bei Kleinkindern* (= *Slaviska institutionen vid Lunds universitet, Slaviska och baltiska studier* 6) (Lund, 1963).

5. By comparing Tables I and II we see that the speech development process in both children did not proceed in the same speed and that the Latvian boy learned to use the negative and affirmative forms earlier than the girl. Nevertheless, the RELATIVE chronology is in general the same in both cases:

(a) first use of 'hide and seek' type game with the corresponding utterances *tutū* (when disappearing) and *ķiķī* (when ıeappearing) at 1;4 (Latvian boy) and 1;3 (Latvian girl),

(b) first use of antonym pairs at 1;6 and 1;10,

(c) greater interest in actions and opposites at 1;7 and 1;11,

(d) numerals 'one' and 'two' (as opposed to 'one' and meaning both 'two' and 'several') at 1;7 and 1;11,

(e) significant increase in number of actively used verbs at 1;9 and 2;0,

(f) increase in adult type negative verbal forms at 1;10 and 2;0,

(g) prefixed verbs in active use at 1;10 and 2;2,

(h) first use of negative pronoun *ne·kas* > [kas] 'nothing, no one' at 1;10 and 2;5,

(i) between 2;0-3;6 incorrect use of Latvian multiple negation still occurred.

Common to the acquisition process of negative utterances in both children was that:

TABLE I

(Latvian boy) [ḿ-m] and [m-ḿ] but mistakes in using the negative and affirmative form.	1;3
first use of 'hide and seek' type game; [m-ḿ] 'yes' + nod : [ḿ-m] 'no' + shake of head; first use of 'empty hand's gesture in the function of verbal phrase: [mȩmmȩ] + gesture 'mummy isn't (there)' but [mȩmmȩ̇] (without gesture) 'mummy is (there)'.	1;4
mistakes in using [jā] 'yes' and [nē] 'no'.	1;5
[nē] | | [dē] 'no' and (less frequently) [jā] 'yes'; but sometimes mistakes in using them; [nav] | | [dav] 'isn't', often together with the 'empty hand's gesture; ANTONYMS (IN PAIRS) IN ACTIVE SPEECH (cf. 'little' — 'big', 'silly' — 'clever', 'full' — 'empty').	1;6
correct use of [jā] 'yes', [nē] 'no' and [nav] 'isn't'; GREATER INTEREST IN ACTIONS AND OPPOSITES; NUMERALS 'one' as OPPOSED TO 'two' or 'several'.	1;7
first use of the negative particle [ne-] in the verb form; [nȩvȩ] = Standard La. *nevar* 'cannot', e.g. [mȩmmȩ va, Daī nȩvȩ] 'mummy can, Dainis cannot'; the common substitute for negative verbal forms at this time was the use of [nē] 'no' in the beginning, at the end or in the middle of the phrase; [nav] 'isn't'.	1;8
SIGNIFICANT INCREASE IN NUMBER OF VERBS; parallel use of negative verbs (in accordance with the adult's practice) and [nē] 'no' + verb, e.g. [Daī ne-baza mute!] 'Dainis did not put into the mouth!' and [Daī nê mute!] 'Dainis no mouth' = 'I did not put into my mouth'.	1;9
INCREASE OF ADULT TYPE NEGATIVE VERBAL FORMS; first use of negative adverb [maz] = *ne·maz* 'not at all'; first use of negative adjective [nitī] = *netīrs* 'dirty'; first use of negative pronoun [kas] = *ne·kas* 'nothing'.	1;10

TABLE II

1;3	(Latvian girl) first use of 'hide and seek' type game.
1;4	dislike, protest by crying; [nâ] = *nav* 'isn't/hasn't'.
1;5	
1;6	often [tutū] 'isn't' in combination with [n̦â:] 'isn't'.
1;7	[n̦ę̂] 'no'; 'isn't' + shake of head; *jā* 'yes' not used.
1;8	[papa ' nàu mę̄ṁmę] = *cepures nav memmei (galvā)* 'hat not-has mummy' = 'mummy has not a hat (on)'; affirmative answer expressed by [l'ab l'ab!] 'good, good!'
1;9	[nē] 'no', [ḿ-m] 'no', shake of head; [tutū] and [nàu] 'isn't' as synonyms, e.g. [nàu mę̄mmę kap 'kapu] and (after a while) [tutū mę̄mmę kap 'kapu:] 'has not mummy typewriter' = 'mummy has not typewriter'; first use of [nē] 'no' with a following verb: [nê al̄ja] 'no go-bye-byes' in answer to the mother's question [vai Sarma negrib iet gulēt ?]
1;10	affirmation: 'good, good!' (often + nod), e.g. [apl̄! l'ab l'ab!] (+ nod) "orange! good good!" = 'yes, I want an orange, please give me!' [nau] 'isn't/hasn't' is often used but not [ir] 'is/has';

(Table I, continued)

PREFIXED VERBS MORE FREQUENT; negative verbs in correspondence with the adult practice.	1;11
first use of negative pronoun [kāc] = *ne·kāds* 'no, none'; [abim divim nàu vaja] lit. 'both two have-not time' = 'neither one nor the other has time', instead of *ne vienam, ne otram nav vaļas*; first use of *nekas* 'nothing' in a phonetic form with the negative particle: [tu nàu · nikā̃] = *tur nav ne·kā* 'there is nothing'; incorrect use of [visiem ·nemaz nesāp] lit. 'all not-at-all not-feel pain' = 'nobody feels pain', instead of *ne·vienam ne·maz nesāp.*	2;0

(Table II, continued)

	FIRST ANTONYMS (IN PAIRS) IN ACTIVE USE, e.g. 'hot' — 'cold', 'sour' — 'sweet'.
1;11	affirmative [jā], although more seldom than [nē] 'no'; [jā] usually together with a nod; correct use of [jā] and [nē]; GREATER INTEREST IN OPPOSITES; NUMERALS 'one' AS OPPOSED TO 'two' or 'several'; [i] 'is/has' and [nav] 'isn't/hasn't' not frequently used; the affirmative and negative gestures dominate; the opposite to [nav] 'isn't' often [bū] 'shall be' (future tense); NO NEGATIVE VERBAL FORMS IN ACCORDANCE WITH THE ADULT PRACTICE; e.g. [pupu guḽ nê] lit. 'shoes sleep no' = 'one cannot sleep with shoes on'.
2;0	INCREASE IN NUMBER OF VERBS; sentences consisting of four words; instead of [jā] sometimes 'good, good!' [nē] + verb as substitute for the adult's negative verbal forms: [ņęñņę nê! ĭt kũla] = *Dainis neiet skolā* 'Dainis no! go school' = 'Dainis does not go to school'; [nav] 'isn't/hasn't' often used in contrasting, as: ['nau aḱī, l'īľī] = *nav karsta, silta* '(the egg) is not hot, (it is) warm'; [màmma tû, męmmę nê tû!] = *Sarma turēs, męmme neturēs* lit. 'Sarma (wants to) hold (the cup), mummy no hold' (the girl wanted to hold it herself).
2;1	first negative verb [' nebu:] (future tense) in active use; mistakes in using [nav] 'isn't/hasn't' instead of [nē] 'no'.
2;2	FIRST USE OF PREFIXED VERBS; INCREASE OF ADULT-TYPE NEGATIVE VERBAL FORMS, e.g. [męm̃mī, neguli! t'el'ī apā!] 'mummy, don't sleep, get up!'; mistakes in using [nav] 'isn't/hasn't' instead of [nē] 'no' and [ne-] 'not', e.g. ['nau t'ãpa!] = *nesāp* '(the foot) does not hurt'
2;2 — 2;4	[nav] still used instead of [nē] and [ne-], e.g. [vê nã miêg, ņęņņę 'naṷ nã miêg] = *(man) vēl nāk miegs, Dainim nenāk miegs* 'I am getting sleepy, Dainis is not getting sleepy'.
2;5	first use of negative pronoun [kas] = *ne·kas* 'nothing'.

---

(1) During the early periods the children made mistakes in using the affirmative and negative forms, i.e. by using the negative gestures or negative utterances when they in fact meant the opposite (as could be judged by the children's actions). Wholly correct use was first achieved (by the boy) at 1;7.

(2) In early stages, the negative gestures or crying in protest were dominant and used either instead of verbal utterances or together with them.

(3) The use of Latvian negation forms *nē* 'no', *ne-* 'not, un-' and *nav* 'is not' was incorrect — they were confused, i.e. one type of these three negations could take the place of the other two. In other words, the children treated these three types 'no', 'not' and 'is not/have not' as three optional variants of the same negative sign. The specialization of their function in accordance with adult practice came somewhat later.

(4) During some earlier periods the negations 'no' and 'is not/ have not' appeared more frequently in the active speech than 'yes'.

(5) Negative verbs and nouns (corresponding to standard Latvian

(Table II, continued)

3;4	[tu nebi 'kata!] = *tur nebij ne·kā* 'there was nothing'; incorrect use of [tu 'abâ naų iêķâ 'kat'a] instead of *ne vienā, ne otrā nav iekšā ne·kā* 'neither in the one, nor in the other (box) is there anything'.
3;7	active use of [viêns] = *ne·viens* 'nobody', e.g. [viņ vaĩ neva ikuôt' 'viênam] 'they can bite nobody any more'.
4;2	[ne·maz] 'not at all'.
4;9	correct use of the type as mentioned 3;4 [męm̃mît, es nevaru teų aduôt ne viênu, ne uõtru] 'mummy, I can't give you neither the one nor the other', cf. c. 4;0 (the affirmative opposite) [eś tikai juõkuôjuôs: abaś eś sûtīšu Vaĩdaî viênâ apluôksnîtê abas!] 'I was only joking: I shall send both to Walda in one envelope, both' — in accordance with the adult practice.

and consisting of *ne-* + verbs in positive form) were learned later than negative and affirmative INTERJECTIONS 'no' and 'yes'.

(6) Both children began to actively use negative verbs with *ne-* only after a greater INTEREST IN ACTIONS had arisen and a significant number of verbs had appeared in their active speech. The connection is clear between increasing interest in actions and use of verbal forms on one hand and the appearance of negative verbal forms on the other hand.

(7) A forerunner of the use of negative sentences was the practice of asking questions and answering them by the child himself as: [te? nê! - te? nê!] 'here? no! here? no!' (already used by the Latvian boy at 1;8).

(8) Before the children had learned to master the negative verbal forms in accordance with the norms of standard Latvian, the negative interjection *nē* 'no' was used as an equivalent for these verbs, cf. standard Latvian sentence *es nebāzu mutē* = (familiar style) *Dainis* (the name of the child) *nebāza mutē* 'I did not put into my mouth' appeared in the speech of the Latvian boy at 1;9 as

[Daĩ ne-baza mute] 'Dainis not-put mouth'

and was repeated as

[Daĩ nê mute] 'Dainis no mouth' (Fig. 2).

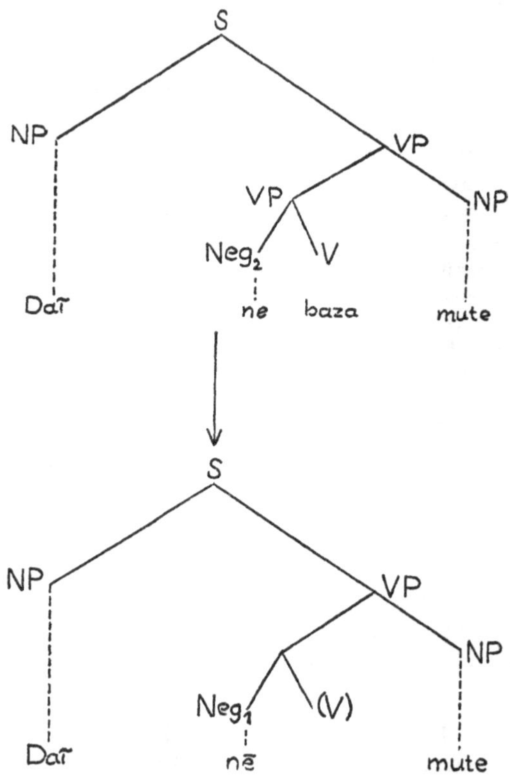

Figure 2.

(9) The next stage in the speech development process was the use of the respective verb in the positive form + *nē* 'no' after it, as in:

['gib nê!] (= standard Latvian: *negrib*) (La. boy 1;10) '(I) don't want!'
[dabu: nê!] (= *nedabū*) 1;10 '(I) don't get' or
[bada nê!] (= *nebadās*) 1;10 '(he) doesn't butt'.

Thus, the difference between the adult's negative verb type and the child's utterance was the different position of the negative element and replacement of the negative prefix 'not' with the negative interjection 'no'.

(10) As the preliminary stage to the correct use of adult constructions, 'no' appeared placed BEFORE the respective positive verb but still not coalescing into one unit with the following verb:

> [Daĩ tĩ nē ʒeś] (= *Dainis tēju nedzers*) 1;10 'Dainis tea no will-drink' = 'I will not drink tea',
> [apu nê dabu] (= *lapu nedabū* = *lapu nevar dabūt*) 'sheet no get' = 'I can't get the sheet'.

(11) The position of this *nē* 'no', used as an equivalent of the negative verb or as a negative sign together with the verb, was thus free in a negative sentence: *nē* could be placed at the beginning as well as in the middle or at the end of the phrase, cf. (La. boy at 1;8):

(a) [nê mẹm̃me t'ip t'ip t'ip] (= *mḷmme neraksta ar mašīnu cipi cipi*) 'no mummy clack! clack! clack!' = 'mummy doesn't type clack! clack! clack!'

(b) [ipupu nệ ĩta!] (= *uz pḷuŋku ipu negribu iet*) 'bathroom no go!' = 'I don't (want to) go to the bathroom'

(c) [ĩta ugu nệ!] (= *pie uguns iet nedrīkst*) 'go fire no!' = 'it is not allowed to go (near) the fire'.

(12) Negative verbs of the adult type with *ne-* became more common during the period when other PREFIXED verbal forms (e.g. those with *pa-*, *no-*, *sa-*) were learned.

(13) The negative verbs as used by the adults were acquired by the children at the same time as the numerals 1, 2 (and 3) were mastered and when 'two' appeared as opposed to 'one', meaning 'more than one' i.e. 'two', 'three' or 'several'.

(14) When negative pronouns and adverbs corresponding to adult forms *ne·viens* 'nobody', *ne·kas* 'nothing', *ne·kad* 'never', *ne·maz* 'not at all' appeared for the first time in the speech of the Latvian children, the phonetic form of the actively used constructions did not contain the (unstressed) negative particle. Thus, the

surface structure of the children's utterances seemed to be a type of simple negation instead of the adult multiple one, for example, the phrase in standard language *nav ne·kas, nē!* 'it's nothing, no!' appeared in the speech of children as: [nav kas, nē!] 'isn't anything, no!'.

As soon as the Latvian boy began to pronounce these negative pronouns and adverbs as disyllabic words with *ne-*, he put the accent on the negative particle, in full analogy with the stressed *ne-* in Latvian negative verbs, nouns and adjectives (but contrary to the adult practice in negative pronouns and adverbs): *·nekā, ·nemaz* — instead of his earlier forms *kā, maz* and in contrast to the adult forms *ne·kā, ne·maz*).

(15) Both Latvian children had learned the negation system of their mother-tongue (in general terms) by the age of 2;0-2;4 years. Nevertheless, they had difficulties with transformations of sentences containing 'all' and 'both' into sentences containing 'nobody', 'neither one nor the other'.

Thus, the opposite to the positive utterance *visiem kokiem* IR *lapas* 'all trees have leaves' appeared in the speech of the children as [visiem kokiem *nav* lapas], where *ir* 'are/have' in the positive sentence was changed into *nav* 'are not/have not' but all other elements of the sentence were left unchanged. In adult speech the transformation must extend to the pronouns as well, the correct form for the negative sense being: *ne·vienam kokam nav lapu* 'no trees have leaves' (lit. 'no-one tree has-not leaves').

Similarly: *visiem sāp* and the corresponding negative *ne·vienam ne·maz nesāp* 'all have pain' and 'nobody has pain' appeared in the speech of the child as: [abiem sāp] 'both have pain' and [abiem nesāp] 'both not-have pain' = 'neither the one nor the other has pain'.

Mistakes of this type first began at the point at which the child had begun to use negative quantifiers like 'nobody', 'nothing' (even if the phonetic (surface) form of them still did not coincide with the adult forms). Competence in handling this type of negative transformation was not reached until the fifth year.

6. The comparison of the Latvian material with facts gathered

by pedolinguists about other languages leads to some conclusions of universal character:

(1) In the pre-speaking stage, the negative reaction is expressed by crying and gestures.

(2) In the speech of young children, negative interjections commonly precede the affirmative ones in time of acquisition.[2]

(3) Affirmative articulate statements (consisting of two or more words) are produced by young children earlier than the negative ones and are used more frequently.

(4) Characteristic for the early stages of active use of affirmative and negative interjections is the confusing of 'yes' and 'no' and the sometimes inconsistent use of them.[3]

(5) Young children do not learn to express negation and affirmation by gestures and by articulate speech at the same time: gesture speech and unarticulated colloquial affirmative and negative forms such as [m-ḿ] and [ḿ-m] seem to be easier to master than articulate speech forms.

The absence of actively used negatives does not necessarily mean that the child is not aware of affirmative and negative SENSE. Children who well understand the negative sentences used by adults but begin talking late, often continue for a long time to use the (negative and affirmative) gestures instead of words.

[2] In the investigations of Karel Ohnesorg (*Fonetická studie o dětské řeči* [1948], p. 70; *Druhá fonetická studie o dětské řeči* [1959], p. 80) the Czech children used *ano* 'yes' earlier than *ne* 'no'. Ruth Hirsch Weir, *Language in the Crib* (1962), pp. 93ff. analyses the vocabulary used in the pre-sleep monologues of an English-speaking boy at 2;6 and finds that *no* and *not* were mentioned frequently but *yes* not at all. Leopold's children began to use 'yes' and 'no' almost at the same time (1;6), cf. *Speech Development of a Bilingual Child* III, 7ff.

[3] W. Leopold, *Speech Development* III, 143 means that "the designation of logical opposites by the same word is a striking special case of extension of meaning. It puzzles the adult that children use the same word for 'up' and 'down', for 'warm' and 'cold', even for 'no' and 'yes' ... designation of opposites by the same word is not unknown in standard languages" and points to Preyer who noticed affirmation instead of denial at 1;11, at 2;4 and 2;7; 'too much' instead of 'too little', 'never' instead of 'always'. I would mention also the different practice in standard languages in using 'yes' and 'no' with the same sense in answers as in Russian: *Vy ne student?* — Net, *ja student* and in English: *Aren't you a student?* — Yes, *I am a student.*

(6) As an important fact from the semantic point of view, the negative sense 'is not' appears to be clearer and more easily grasped by young children in situations where something already apparent suddenly disappears, e.g. when a car passes, an object is engulfed by the waves or something is not longer found at the same place as before.

(7) At some stage young children treat all negative forms used in the speech of adults as variants of the same negative sign and use one of them also as equivalent for the others. Thus, in the speech of Russian children, the independent negative element *net* 'no; is not' appears in all cases, even where the adult form would have only the single negative element *ni*, e.g. instead of *ni karmí* 'don't feed' the child says *net kalmí*.

Similarly the Latvian children can substitute *ne-* 'not' with *nē* 'no' or *nav* 'is not'. Further specialization of the negative forms in accordance with adult practice comes somewhat later.

(8) Individual affirmative and negative forms may be intended by the children themselves by analyzing and misunderstanding the sense of some words or word groups in the corpus of adult speech, cf. 'gone to work', 'all gone', etc. as the usual way of indicating the absence of something; or 'good, good' used as an affirmative interjection instead of 'yes'.

(9) The degree of mental development reached by children beginning active use of negative and positive opposition, coincides with the stage, when antonym word pairs (such as 'little' — 'big', 'full' — 'empty', 'clever' — 'silly') appear in their speech, and when they master numerals 'two' (meaning 'two', 'several' or 'many') as opposed to 'one', and when 'three' as the third step in the numeral system 'one', 'two', 'three' first appears. At age of two, children are often concerned with paired associations, and as a result of this, contradictory utterances are frequent in their speech during this period, for example, *not the yellow blanket, the white.*

(10) In learning languages in which negative verbs are formed by negative prefix (as in Russian, Czech, Latvian), young children acquire these negative verb forms at the same time as other pre-

fixed verbs (with prefixes *pa-*, *no-*, *sa-* etc.) become active in their speech.

(11) Before the active use of negative verbs formed with negative prefixes — as a natural pieliminary in the language acquisition process — the negative interjection 'no' — and somewhat later this interjection 'no' + verb replaces the negative verb form of adult type, for example, *No I see truck* as an answer to the question *Did you see the truck*? or *no, go* instead of *don't go*.

(12) Linguistic systems for expressing negative sense are not equally intricate in all languages and take shorter or longer time for young children, depending on their respective mother-tongues. At the age when e.g. Latvian children have already acquired negative verb forms in accordance with the adult practice, English-speaking children have usually still not learned the transformations with *do* in negative utterances in correspondance with standard English use.[4]

(13) Transformations appear relatively late in child language. During an early period, the transformation of affirmative statements into negative ones is incomplete. As a natural consequence of this, some elements remain unchanged in negative sentences in the speech of children, and the result may be simple or multiple negation, differing from the adult practice. Thus analogue, building on the basis of the affirmative sentence *I saw nothing* may result in *I didn't see nothing*, or *I saw something* may be transformed into *I didn't see something*, where only one member in the affirmative phrase, namely the verb, is replaced by the corresponding negative form, whereas the words *nothing* and *something* are left unchanged as in the basic (affirmative) sentence.

[4] Ruth Hirsch Weir, 1962: 86, 92, 144: "The standard English 'do' constructions of sentences have not yet been learned by the child. Although negative imperatives using 'do' are quite common in the corpus and some declarative constructions are used as well, there is only a single example of a 'do' construction in an interrogative without a question word, nongrammatical at that: *Did Penny wants water*" (p. 92). "*Don't jump* or *No, go*, the second type being less frequent" (p. 86). Cf. also Roger Brown and Ursula Bellugi, "Three Processes in the Child's Acquisition of Syntax", *New Directions in the Study of Language*, ed. by Eric H. Lenneberg (1964), 135.

The Latvian example cited above: (affirmative) *visiem kokiem* ir *lapas* 'all trees have leaves' and (negative) *visiem kokiem nav lapas* 'no trees have leaves' (lit. 'all trees have-not leaves') instead of standard Latvian *ne·vienam kokam nav lapu* 'no trees have leaves' (lit. 'no-one tree has-not leaves') shows clearly that the young child had not yet mastered the rule:

neg + *viss* 'all' + noun → *ne·viens* 'no one' + noun.

The complete transformation is not possible until the Latvian child has learned to master this additional transformation rule, namely, that not only must 'have' be replaced by 'have not', but also, accordingly to Latvian language rules, negation must be applied to all quantifiers in the sentence and not to just a single element. In addition, the child must know that the nominative form *(lapas)* simultaneously with the negative transformation, must be changed into the genitive *(lapu)*.[5]

(14) In the transformation process the affirmative sentence is usually the basic sentence form for analogue building. For example, in standard Russian transitive verbs in the negative usually take the genitive case instead of the accusative *(ja ne pišu piśma)* 'I am not writing [of] a letter') and impersonal phrases formed with *net* and with nouns in the genitive case are used frequently. According to Gvozdev,[6] Russian children go through a stage of using the accusative and nominative instead of genitive also in negative sentences, as they do in the corresponding positive sen-

[5] In my opinion, Paul Kiparsky is wrong in generalizing that "some relationship between 'substandard' *neg*-incorporation and 'standard' *neg*-incorporation may be responsible for the fact that the former is the natural predecessor of the latter in the development of a child's linguistic system and also the natural result of the latter by linguistic change" ("Linguistic Universals and Linguistic Change", *Universals in Linguistic Theory*, ed. by Emmon Bach and Robert T. Harms, 170-202; 194). His statement "... instead of producing the 'normal' sentences like *I saw nothing*, *Nobody likes me* he at first [! VR] comes out with *I didn't see nothing*, *Nobody don't like me*" may be true for English-speaking children but comparative material from other languages shows that this statement cannot be universal.

[6] A. N. Gvozdev, *Voprosy izučenija detskoj reči* (Moskva, 1961), p. 27-28. Cf. also Dan I. Slobin, "The Acquisition of Russian as a Native Language", *The Genesis of Language*, ed. by Frank Smith and George A. Miller, 133.

tences with the same verbs. Heie, too, the explanation lies in analogue building with the affirmative statement as the basis and in incompletely mastered transformation rules.

(15) The emergence of affirmation and negation in child language cannot be studied as an ISOLATED phenomenon but only as a part of the whole language-learning process.

Special attention needs to be paid to such factors as the point at which new lexical and grammatical categories appear in active speech, e.g. a significant number of actively used verbs, which opens the way for the use of negative verb forms, or the acquisition of a new morpheme, e.g. verbal prefixes, which (in the case of Latvian) facilitates the active use of verbs formed by negative prefix *ne-*. The first use of negative quantifiers often introduces a period of simple or multiple negation differing from the adult practice, and so on.

In conclusion, the phonemic system of the particular language must also be taken into account in the emergence of special syntactic models. In languages with negative pronouns and adverbs pronounced with unstressed negative prefix, these forms may first appear in the speech of children without the negative element: [nav kas] (= standard Latvian *nav ne·kas*) 'it's nothing' or [net kavó] (= standard Russian *net nika·vo*) 'there is nobody'. Thus, the surface structure of such negative utterances used by the children resembles sentences with simple negation instead of those with multiple negations, as in the standard mother-tongue. This phenomenon is not typical for Czech children learning their native language, as the corresponding negative pronouns and adverbs in Czech have the same stress on the first syllable as negative nouns and adjectives.[7]

The emergence of negative sentences must be discussed in intimate connection with the syntactic development in general (such as the length and structure of sentences actively used by the child); on the other hand, the appearance of negative forms can expand the dimensions of a phrase, as the first two and three-

[7] Cf. Karel Ohnesorg, *op. cit.*, (Praha 1948), 58.

word sentences in the speech of young children often consist of one or two-word sentences + a negative element.

REFERENCES

Brown, Roger, and Bellugi Ursula,
1964 "Three Processes in the Child's Acquisition of Syntax", *New Directions in the Study of Language*, ed. by Eric H. Lenneberg, 131-161.

Brun, Theodore,
1969 *The International Dictionary of Sign Language* (London).

Gvozdev, A. N.,
1961 *Voprosy izučenija detskoj reči* (Moskva).

Kiparsky, Paul,
"Linguistic Universals and Linguistic Change", *Universals in Linguistic Theory*, ed. by Emmon Bach and Robert T. Harms), 170-202.

Klima, Edward S.,
"Negation in English", *The Structure of Language: Readings in the Philosophy of Language*, ed. by Jerry A. Fodor and Jerrold J. Katz, 246-323.

Leopold, W. F.,
1939-1949 *Speech Development of a Bilingual Child* 1-4 (Evanston).

McNeill, David,
1966 "Developmental Psycholinguistics", *The Genesis of Language*, ed. by Frank Smith and George A. Miller, 15-84.

Ohnesorg, Karel,
1948 *Fonetická studie o dětské řeči* (Prague).
1959 *Druhá fonetická studie o dětské řeči* (Brno).

Rūķe-Draviņa, Velta,
1959 "Zur Entstehung der Flexion in der Kindersprache", *International Journal of Slavic Linguistics and Poetics* 1:2, 203-222.
1963 *Zur Sprachentwicklung bei Kleinkindern* (Lund).

Slobin, Dan I.,
1966a "Abstracts of Soviet Studies of Child Language", *The Genesis of Language*, ed. by Frank Smith and George A. Miller, 363-386.
1966b "Comments on 'Developmental Psycholinguistics'', *The Genesis of Language*, ed. by Frank Smith and George A. Miller, 85-91.

Waismann, F.,
1965 *The Principles of Linguistic Philosophy* (New York).

Weir, Ruth Hirsch,
1962 *Language in the Crib* (The Hague).

DISCUSSION:

PAVLOVIĆ: Pour la psychologie linguistique la négation est de caractère bivalent, (a) constatation (faits de situation), (b) attitude du sujet parlant (avec les verbes volutifs). Les observations du langage de l'enfant se prêtent à faire remarquer les constatations précieuses indiquant formulations des types de négation d'après les types de langues.

MIKEŠ: Important points for further investigations:
— the development of negative form in the framework of the whole development complex significant to that age,
— negation and SPS.
Some examples from material Mikeš-Vlahović: first neg. forms in *neím*, *nema*; examples of pre-transformational forms.

KAPER: Welcher Art wird der Fehler in: *yes* - *no*? (z.B. *yes* statt *no* und nicht umgekehrt?). Auch *(n)irgendwo*, vgl. *(n)ever* durcheinander?
Welcher Zusammenhang zwischen Negation/Affirmation und Zahlwörtern?

SKOUPIL: Die negative Form soll an der Grundlage des Sprachsystems erforscht werden, weil in manchen Sprachen es einen hauptsächlichen Unterschied gibt zwischen Negationsform mit Negativpräpositionen (welche konträre Gegensätze ausdrückt) und Negationsform mit einer anderen Wortwurzel (welche kontradiktorische Gegensätze ausdrückt).

A METHOD OF DESCRIPTIVE ANALYSIS OF CHILD ACTIVITY IN THE PERIOD OF LANGUAGE ACQUISITION

GRACE WALES SHUGAR

(Warszawa)

1. There are two contrasting ways in which language acquisition has been studied, both contributing to knowledge and theory in this domain. One is to investigate language development in isolation from other aspects of activity (the latter, at most, taken into account as non-linguistic context). This approach underlies most of the vast literature on the subject[1] including current research on grammar acquisition in developmental psycholinguistics.[2] The other way to approach the study of language development is in a strict relation to that of total activity. This relation can be studied according to various aspects. A.R. Luria[3] and other Soviet investigators have drawn attention to the function of verbal activity as regulator of other forms of activity. We propose here to approach this relation in the following manner. Our basic tenet is that verbal and non-verbal behavior are in themselves but different facets of a broader process. This process is such that it takes on many diverse forms, yet always possesses one basic characteristic, namely, directionality. We accept that the form of this

[1] Cf. for example D. McCarthy, "*Language Development in Children*", in Carmichael, *Manual of Child Psychology*, 2nd ed. (New York, 1954).
[2] Cf. for example U. Bellugi and R. Brown (eds.), *The Acquisition of Language* (= *Monogr.Soc.Res.Child Development*, 29:1) (1964); F. Smith and G. A. Miller (eds), *The Genesis of Language: A Psycholinguistic Approach* (Cambridge, Mass., MIT Press, 1966).
[3] Cf. A. R. Luria, "The Directive Function of Speech in Development and Dissolution", *Word* 15:3 (1959).

process specific to man, and which takes shape gradually during ontogenesis, is 'directed action'. This concept may be defined as that process which, by virtue of its essential feature of directedness, acquires structure appropriately to conditions so that a course leading to an end-result is maintained.[4] Following this view — one which is more and more widely held in psychology today and is extensively elaborated in Polish psychology[5] — we take for granted that child activity, possessing this feature of directionality from the start, progressively becomes directed to the attainment of something, in a gradual course of development during which all available mechanisms — both verbal and non-verbal — are set into operation toward this end.

If such a postulate is accepted, clearly a fundamental task facing the developmental psychologist interested in language acquisition is to gain knowledge of the changes occurring with age in the domain of child action and its programming and implementing mechanisms.

In the hope of contributing to this important task, this paper presents a proposal for a method to analyze directed activity in the child, which permits both a qualitative and quantitative description of developmental changes in molar activity structure, and also enables us to grasp changes with age in the structure and function of utterances. This method should also allow for a quantitative control of reliability of measurements and analyses performed.

First — I shall report on the method, then — present some early results as an illustration of the kind of finding we can expect from the use of this method.

2. In order to accomplish the task I have just outlined, a method must do four things:

[4] T. Tomaszewski, *Wstęp do Psychologii* [Introduction to Psychology] (Warszawa, PWN, 1967), p. 139.

[5] Cf. note 4; also M. Maruszewski, J. Reykowski, and T. Tomaszewski, *Psychologia jako nauka o człowieku* [Psychology as the study of Man] (Warszawa, Książka i Wiedza, 1967). (Among the best known precursors of this approach is E. C. Tolman).

(1) It must be able to discover the single manifestations of directed activity — the units of molar action;
(2) It must elaborate the criterion of directedness for application to empiric data. In its most general terms, the criterion is that two successive units differ from each other by a change in the direction of action; that is, we must look for differences in direction to discover our units;
(3) It must allow for isolating units in relation to their immediate situation in the on-going stream of events. The concept 'immediate situation' is understood as the point of juncture between external and internal conditions which permits the action to occur and determines its direction and shape. An action unit in relation to its immediate situation is called an episode;
(4) The method must lead to isolating units in such a way as to maintain their complex interrelations within the stream of behavior.

The present method has been inspired by the work of Barker and his collaborators described in his book *Midwest and Its Children*.[6] It is an adaptation of Barker's method for the more detailed analysis of child activity between the ages of 1½ and 3 years — period of most rapid language growth.

We proceeded as follows:

Data was collected by tape-recorded observation of two children's activity in natural surroundings with verbatim speech, including all other relevant activity and speech. Observations on each child separately were made at two-week intervals over the space of one year. The data was transcribed in total on typed sheets, about 40-60 per observation. These are the raw materials on which the work of episoding is performed.

To get an idea of what episoded material looks like, it suffices to look at two work-sheets from the typed observation record (p. 246, 247). On the right side is the raw data, line by line in temporal order of sequence. On the left these data are chunked into

[6] R. G. Barker and H. F. Wright, *Midwest and Its Children: The Psychological Ecology of an American Town* (New York, Harper and Row, 1954). Cf. also R. G. Barker *The Stream of Behavior: Explorations of its Structure and Content* (New York-London, Appleton-Century-Crofts, 1963).

units or episodes bracketed off and numbered. Notice the overlap of episodes. They occur simultaneously as well as consecutively. Some are longer and enclose others. Some are short and are enclosed, or nested, within their embracing episodes. Notice also how utterances occur here and there throughout the data. These two work-sheets together make up a fragment of the activity of Małgosia, aged 2 years, 6 months, 22 days, during about 5 minutes in time.

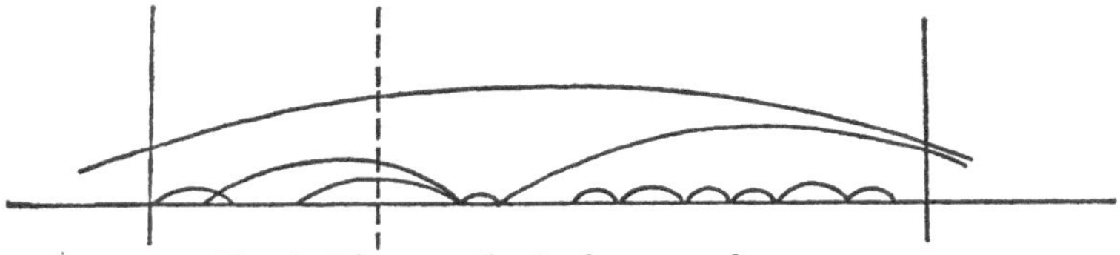

Fig. 1. Diagram of episode pattern from excerpt of observation of Małgosia 2;6,22

Fig. 1 shows the structure of these episodes along their temporal continuum. If we take one moment in time — at the vertical dotted line cross-cutting the data — we could describe Malgosia's three conjoined activities as follows:

While descending the stairs in the newly-mastered adult mode of alternate steps, which takes her along toward an anticipated goal (the playground), Małgosia ventures upon a new kind of action, executed verbally, namely, chatting about a non-present topic — the whereabouts of the garage.

What has been said has served to illustrate the method of episoding and the type of result obtainable.

Now we will speak of the episoder's task and the nature of his equipment. His task is painstaking and long, for he must manage to follow the child in the child's own footsteps, not in his own, keeping to the child's perspective and ignoring his own. For this he needs two kinds of equipment.

One kind is a systematized form of present knowledge of child action, which we have derived largely from the work of Piaget[7]

[7] J. Piaget, *La naissance de l'intelligence chez l'enfant* (Neuchâtel, Delachaux et Niestlé, 1936).

WORK-SHEET (1)

6 *Sample Excerpt from Observation Record XXV. Małgosia* 2;6;22

10.20 a.m. Grandmother has just started off down the 3 flights of stairs

122 C

Małgosia is standing just outside the open door of the apartment,
on the landing, one hand on the handle of go-cart.
E is collecting her handbag and talking briefly with Pani Teresa

niesie | nniesie babcia | niesie
"niesie - (nie)niesie babcia — niesie" (voice fading low)

E: "Co?" now to M, whose eyes are fixed on E

123 C

Pani Teresa breaks off her talk to E to say: "No daj cioci rękę"
while E has picked up the go-cart under her arm

M slips her right hand into E's waiting one,
and looks forward down the flight of stairs,
With a big fling of the left foot and grasping the bannister with
left hand she descends to the next tread,
then with a circular fling of the right, steps down heavily to the next.
Pani Teresa, watching from the doorway, comments: "Dobrze", and closes door
behind her.

124 C

M's eyes are fixed on E's feet.

We go down, more or less rhythmically, all steps alternately, and
in silence for about 20″.
After turning on the landing to the 2nd flight,

125 E

M enquires suddenly gdziewaśtat / "gdzie wasz warsztat?" eyes forward on window looking out on street

Grandmother from lower flight answers: "Gdzie babcia, no zgubił".

gdziewarśtat
"gdzie warsztat" interrupting with strong stress on 2nd syllable

Grandmother: "Nie rozumiem"
E: "Gdzie warsztat?" checking her comprehension.

tak / "tak" briskly, and adds gdzie| tak | ipżyjeżdzajom / "gdzie - tak - i przyjeżdżają"

E: "Który warsztat?" (recalling an earlier conversation this day)

ten | odsamochodu | cielepelują
"ten - od samochodu/ów co/gdzie reperują"

WORK-SHEET (2)

E: "Więc to jest warsztat daleko —"
A moment's pause as we turn the landing onto, last flight of stairs,
then M pursues the topic with: tak
E: "... tam gdzieś w Warszawie" (in tone as if recounting a tale)

wwarśawy — "w Warszawy" and echoing after E gdzyświwarśaw — "gdzieś w Warszaw..."

but she breaks off, now bouncing down faster the last few steps
drops E's hand,
and runs out the door,
126 Ed past a broom leaned against the outside wall

patćzostałapaninadwoźmiatły — "patrz została pani na dworzu miotła" with a humorous flavour

127 C and with both hands grasps the handle of the gocart set down by E,
pushing it at once along the pavement to the right.
The gocart wheels squeak.
Grmo grimaces in distress to E who exclaims also: "Och, jak to piszczy ... tak"
M pushes, ahead, speeding up a bit which produces a faster rhythm of squeaks.
128 Ed Her eyes are fixed downward as if on her moving feet.
To the left small boys' voices rise from the sandbox in the yard
and M casts a glance toward them.
At a fork in the pavement M makes a right turn as if automatically
but Grmo calls: "Nie, na lewo",
and M immediately without stopping swerves her gocart about and follows grandmother to the left.
129 C For about 20″ M goes her way in silence, ignoring a barking dog

then piścitenwóziek piściten (louder) wóziek piści ten — "piszczy ten wózek, piszczy ten wózek - piszczy ten" takes a breath

wóźek piścitenwoźek — "wózek - piszczy ten wózek" with humor now in last segment

Grmo murmurs some reply but M interrupts with piścitenwóziek — "piszczy ten wózek"
loudly, as if to the deaf
E (feeling herself the addressee) hastens to reply: "No bardzo, ... nawet ... bardzo".

and of Szuman.[8] This takes the form of a representation of what the child is capable of doing with respect to the information accessible to him. It is, in effect, a working model of an action, which can be matched against each instance of action in its immediate conditions. The purpose of this auxiliary model is to account for the empiric facts and lead to the discovery of inner patterns of action. Without discussing it at length, we will say only that it is comprised of two types of constituents: one are its elements such as agent, object, receiver, the action-proper (corresponding to the verb in a sentence), means, effect, and others, some of which are necessarily present and some not.

A necessary constituent considered of decisive importance is RECEIVER. An action can have more than one receiver (e.g., the intended), but always has at least one, which is the agent. Otherwise stated, the child always experiences his own action, in accordance with the feedback principle. In the flow of behavior, constituent elements constantly change, and transform one into another. In certain developmental stages some take precedence over others. Our analysis requires that we recognize the particular constituents of activity in their changing constellations.

The other type of constituent is that of stages through which the action flows along its course to an end-state. It is accepted for this model that there are three basic stages, namely, a reflexive-reactive stage, an approaching (or withdrawing) stage; and an executive stage during which the action is realized leading to an end-state differing from the initial state by the effect of the action. In addition there are auxiliary stages which emerge here and there, embedded within the main stages. The principal ones may be labelled as follows: preparing the action, coping with difficult conditions, evaluating the action, improving on performance. These stages seem somehow to emerge developmentally, and presumably integrate themselves eventually into some relatively balanced and mature action scheme as terminal product.

8 S. Szuman, *Rola działania w rozwoju umysłowym małego dziecka* [Role of Directed Activity in Mental Development of the Child] (Wrocław, Zakład Im. Ossolińskich, 1955).

We should add that the above-described model emerged in the course of examination of the observation records of the two children in this investigation. It forms a working hypothesis, used so far only for the episode analysis of the first and last of the two series.

In addition to such a conceptual framework — which serves mainly to organize his expectancies — the episoder requires something more. To proceed step-by-step through the thicket of data, he needs a set of working tools for the interpretation of the criterion of directionality. These tools consist of an explicit set of rules of procedure which when applied should lead to uniform interpretations of the data and uniform results. Time does not allow for a demonstration of how these rules work. In a word, they instruct on how to find and use the signposts in the data that mark s'arts, stops, turns, interruptions, resumptions. Not only do they serve to identify the units but also indicate how to link them together by bracketing.

Once processed into episodes, the data is in a form where it can be analysed further in various ways, for various purposes.

3. I shall now report some general findings on the structure of episodes, and then deal with an analysis of more particular interest to this audience.

To date, 4 observations, each of 1½ hours' length, have been analyzed into episodes. These are the first and last in the series conducted on two children, with one year's interval between. At time 1, the children were aged 1; 7; 9 and 2; 0;13, when they were beginning to make two-word combinations in utterances.

In Table I we find that the average length of an episode is very short — about 20″ long in all 4 sets of observations. Speech participates throughout — one to two utterances occur in each episode on the average. Change with age is in utterance length. Judged in number of morphemes, utterance length more than doubles for both Małgosia and Mikołaj.

Now look at the ways in which episodes interlink and overlap, as shown in Fig. 2.

TABLE I

Means of episode length, number of utterances per episode, and utterance length in morphemes, pertaining to two children aged 1;7,9 *and* 2;0,13 *at Time* 1, *and one year older at Time* 2.

	Małgosia		Mikołaj	
	T_1	T_2	T_1	T_2
Mean length of episode	20″	20″	17″	16″
Mean no. utterances/episode	1.4	1.3	2.1	1.2
Mean no. morphemes/utterance	1.5	3.9	2.1	5.1

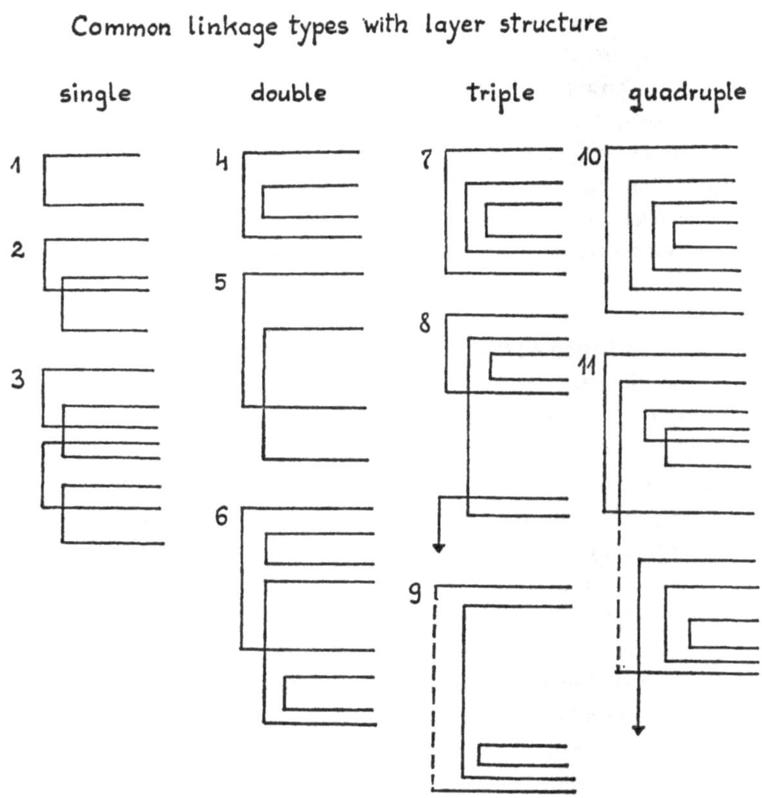

Fig. 2.

A number of general statements can be made on the basis of our findings. We see here a progression from simple to complex structures as the direction of change with age. The isolated episode (item 1), even at the start, is rare and tends to disappear with age. The most common connection between episodes is the interlinking connection (items 2 and 3) and in the more complex structures as well (items 5, 6, 8 and 11). This signifies that some common constituent exists in the actions of the episodes — a thread of continuity, or it may also mean their temporal overlap. About 60% intersect thus with others.

Another dominant kind of linkage is the enclosing-enclosed type as you see in items 4, 6 and all the following. The numbers of enclosing episodes do not alter much with age — remaining around 15 to 20%. Enclosed episodes, however, increase from about one third to two thirds.

A further word about the enclosed episode. It was found that this type differed in functional relation to the enclosing episode. One kind feeds into, adds to or varies the enclosing episode, while the other kind diverts, detracts and leads away from it. The former kind shows a startling jump for both children, a fourfold increase over the year.

Finally, there is the layer structure. More and more layers emerge with age. This means that fewer episodes occur on the single layer, more and more on the second and third.

These changes may be interpreted as indicating integrative processes whereby many co-occurring or consecutive actions get coordinated into larger, more complex actions in a hierarchical organization.

4. How do utterances participate in action units described as episodes? More specifically, let us ask, during what stage of the action does the child say something? I would like to report now, in conclusion, on a pilot study of the distribution of utterances according to the stages of the action. This study was limited to one continuous string of 48 episodes containing 90 utterances, a fragment from Małgosia, aged 2;6,22.

The method used in this breakdown is as follows:

Places of occurrence are defined according to the stage structure of the action of our working model, i.e., three basic stages and the various auxiliary stages.

The analytic task is to diagnose at what stage of the given action the child utters something. The utterance content is treated only as one source of this required information.

The procedure is, first, to read the name of the episode giving its content, then to read the data, diagnose the stage when the utterance occurs and make a type/token count.

Owing to the layer structure of the episoded data, the utterance must be related to each of the on-going episodes. That is to say, the utterance falls into one stage in one action in course, and often into another stage in another, parallel action. This will be clearer when we see the results.

It was found that, in relation to the first or immediate layer, over 70% of the utterances occurred during the executive stages, i.e., the realization of the action, and 18% occurred in auxiliary stages. Of these the most important was the stage called by us: "Coping with difficult conditions". Only a few — 4% — were irrelevant to this layer. The remainder could not be judged.

Now concerning the second layer of longer-term, enclosing and superordinate actions, it was found that much fewer occurred during the realization itself of these actions, i.e., 33%, whereas much more fell into the auxiliary stages — 45%. Of these the large majoıity occurred in the stage called: "preparatory, to-the-action". A much larger number were irrelevant to the enclosing action — 20%. The remainder could not be judged.

Concerning the third layer, the more remote enclosing actions, only 10% of the utterances were related to their realization stage, whereas 53% were related to one auxiliary stage, "preparatory-to-the-action". Irrelevant utterances were 37%.

Only 3 utterances occurred in relation to a fourth layer, all of which, as well, occurred in the preparatory auxiliary stage.

Another finding from this exploratory study was that, of the 49 episodes making up this string, 2 of them could be called verbal episodes *per se*. This means that the constituent elements of the action —apart from agent-receiver (or interlocutors) —were remote in time and place, and that the entire course of the episode was realized in the verbal mode. The names of these episodes were: "Joining in conversation about whereabouts of grandmother", and "Making conversation about a recalled incident involving a teddy bear". 10 utterances comprised these two episodes.

The above illustrates the kind of information we may anticipate as obtainable by such an analysis. It is one of the possible range of analyses that can be carried out on episoded material, one which may conceivably contribute to a clarification of the relation between language and action.

ON LINGUISTIC SENSE FROM THE POINT OF VIEW OF THE REFLEX THEORY

MILOŠ SOVÁK

(Praha)*

The period of meaningful sounds plays an important role in the development of speech and language. The perception, retention and recall of meaningful sounds is called acoustic gnosis (eugnosia). If verbal utterances be used (as meaningful sounds) the verbal acoustic gnosis (word eugnosia), i.e. perception, retention and recall of verbal sounds (or speech patterns) is functioning.

The function of the word acoustic eugnosia is performed on the level of the first signal system, i.e. on the level of concrete thinking. At the lower stages of language development verbal sounds are associated with concrete objects, persons and/or life situations.

By means of the imitation drive the child transfers certain verbal sounds, originally connected with certain objects, to similar (analogous) objects as well. This very important phenomenon is called transfer (analogization). The function of transfer in children is very, sometimes surprisingly precise. Through the developmental period of transfer the generalisation and the differentiation of terms may be achieved.

By means of the ability to transfer, i.e. *per analogiam*, the child also acquires grammatical forms. In the case of developmental agrammatism the word forms made by the child in precise analogy do not respect the exceptions in grammatical rules. Later on developmental agrammatism diminishes and comes to be overdone

* Charles University, Faculty of Education, Dpt. of Special Education.

through linguistic experience based on spontaneous correction by the imitation of the language of other people.

The unconscious reflex ability to transfer helps the child to overcome difficulties in acquiring the language and supports the grown-up in the use of the right term and the proper grammatical form.

Because of the transfer, the newly acquired words and speech patterns are then compared with the existing vocabulary of the individual. This process of comparison involves the identification and/or correction of word-forms.

As another decisive factor in choosing the proper form of words verbal gnosis must be taken into account. Acoustic verbal gnosis is supported by kinesthetic and also — esp. by literate people — by visual gnosis.

From the point of view of the reflex theory the function of transfer, performed in the domain of verbal (acoustic, kinesthetic, or visual) eugnosia represents the reflex basis for so called linguistic sense (skill, sense of language, *Sprachgefühl*), i.e. for the unconscious ability to choose and use the right form of words and speech patterns.

The performance of the linguistic sense depends on intellectual capacity, the individual's linguistic experience and level of education, and on many other psychological factors.

DISCUSSION:

SLAMA-CAZACU: The problem of transfer (in connection with interference) is of interest in the psychology of learning, and in present times it has become a topic which is much spoken in 'contrastive linguistics'. Do you see any connection between the concept of 'transfer' — in the sense you have spoken about — and that one implied in the process of foreign-language learning/teaching?

ANSWER: I was interested in the concept of transfer on the physiological level.

DAS GEFÜHL UND DIE SPRACHE

ÜBER DIE BEDEUTUNG DER AFFEKTE UND EMOTIONEN FÜR DIE ENTWICKLUNG UND BENUTZUNG DER SPRACHE

MARIE VAJSHAJTLOVÁ

(Praha)

Die Entwicklung der Sprache beim Kinde ist mit dem Gefühlsleben eng verbunden. Vom Anfang an wird die Sprache und Ihre Entwicklung durch Gefühle (Affekte und später Emotionen) durchgedrungen und beeinflusst. Mit dem Reifen der sensitiven Nervenbahnen — in Verbindung mit der *Formatio reticularis* — kommt die Gefühlskomponente schon in den ersten Sprachperioden zur Geltung. Schon in den Stimmäusserungen der Säuglinge spiegeln sich die primitiven Lust- oder Unlustgefühle wider.

In der Sováks reflexologischen Auffassung[1] der Sprachentwicklung wird die primäre Rolle des Gefühls eindeutig betont. Mit dem Reifen der Thalamuskerne kommen die Affekte auch in den Stimmäusserungen — auf Grund der vegetativen oder instinktiven Bedürfnisse — zum Ausdruck.

Im Nachahmungstrieb, das heisst in den sensomotorischen Reflexen, kann ein deutlicher Einfluss der Gefühle in der Stimme und in den Sprachanfängen festgestellt werden. Die unbewusste reflektorische Nachahmung der Sprachlaute wird durch eine positive Gefühlseinstellung zu den anderen Leuten unterstützt.

Als eine sehr wichtige Etappe der Sprachentwicklung müssen

[1] Becker, Klaus-Peter, *Lehrbuch der Logopädie*, Berlin 1971. — S. auch ebenda S.254.

die Modulationsfaktoren, d.h. die Melodie, die Intensität, die Quantität, der Akzent und der Rhytmus, betrachtet werden. Reflexologisch handelt es sich um eine Teilfunktion der Anfangsstufen der höheren Nerventätigkeit, die mit den archaischen Strukturen der Grosshirnrinde (sogenannten *Allocortex*) zusammenhängt.

Die Anwendung der Modulationsfaktoren im Sprechen wird jedenfalls durch Gefühlselemente untermalt. Dadurch kann auch der Sinn des Gesprochenes beeinflusst, ergänzt ja sogar geändert werden. Deshalb kann man die Modulationsfaktoren als eine grundlegende Komponente der gemeinsamen sprachlichen Komunikation betrachten.

Es ist bekannt, dass durch diese das Kind fähig ist zu äussern, ob es etwas wünscht oder ablehnt, ob es etwas lieb oder unlieb hat, ob es befriedigt oder unbefriedigt ist, ob es zornig oder zufrieden ist und so weiter. Das bedeutet, dass mittels der Modulationsfaktoren die elementaren sozialen Beziehungen zum Ausdruk kommen können. Mit Recht behauptet Ohnesorg,[2] dass "die Melodie das erste Funktionselement in der Kindersprache auftritt".

Die Funktionen vom *Allocortex*, die in der Phylogenese mit dem Geruchsinn verbunden sind, stellen nach Piľha[3] die Vorstufe des sogenannten intuitiven Intellekts dar. Infolgedessen können auch die Modulationsfaktoren, die eine Teilfunktion vom *Allocortex* darstellen, als eine Vorstufe der interindividuellen sprachlichen Kommunikation betrachtet werden.

In der weiteren Entwicklung der Sprache kommen ausser den Affekten auch die höheren Gefühle, d.h. Emotionen, zur Geltung. Man unterscheidet niedrigere, d.h. situative, und höhere, d.h. ethische, estetische, morale, religiöse und so weiter Emotionen.

Die niedrigeren Emotionen, zum Beispiel Freude, Scham, Unsicherheit und ähnliches, die sich an der Ebene des ersten Signalsystems abspielen, spiegeln sich, auf Grund der individuellen Erfahrungen, in der Stimme und in der Art des Sprechens wider.

[2] Ohnesorg, Karel, *Fonetická studie o dětské řeči* [Etude phonétique sur le langage enfantin], Praha 1948, S.61 n. 178.

[3] Piľha, V., *Alokortex člověka* [Allokortex des Menschen], Praha 1959.

Die höheren Emotionen, an der Ebene des zweiten Signalsystems, beeinflussen die Sprechart und Wortanwendung der Reife des Geistes und dem Intellektsniveau entsprechend.

Die erwähnten Etappen der Gefühlsentwicklung bleiben in der Struktur der Persönlichkeit stecken, so dass die unteren Etappen durch die höheren ergänzt, vervollkommnet und gegeneinander modifiziert werden. Z.B. die Modulationsfaktoren, die als niedrigere Formen des Gefühls gelten, können auch in den höheren Schichten der Kommunikation, in der Deklamationskunst deutlich auftreten und betont werden.

Nach Sováks Auffassung der Reflextheorie, die als methodologisches Prinzip in der Physiologie und Pathologie der Sprache angewandt wird, alles was man tut, wie man sich verhielt und spricht, hängt mit dem Gefühlsleben lebenslang zusammen.

Durch die Erziehung der Affekte und Emotionen wird die Grundlage der Soziabilität, und dadurch der interindividuellen sprachlichen Kommunikation, vorbereitet.

DISCUSSION:

JANOTA: Es ist erfreulich, daß während der Konferenz auch diese sehr wichtige Frage behandelt wurde. Zum Beispiel in der Phonetik gehen wir oft vom sogenannten 'neutralen' Satz aus, aber wir wissen sehr gut, daß solche Sätze im Sprechprozess sehr selten vorkommen, auch in der Sprache der Erwachsenen. In der Kindersprache ist dies noch in größerem Maße der Fall.

SIERTSEMA: Intonation is not only an emotional feature as is suggested in the extract on p. 226*. Intonation patterns are systemized per language and must be learned just as the phonemes and the grammar must be learned. One and the same intonation pattern can have very different meaning-modifying functions depending on the word-material it is 'superposed' on.

SOVÁK: Ich erlaube mir zu wiederholen, was im Vortrag schon betont wurde, d.h. die enge Verbundenheit der Modulationsfaktoren mit dem Gefühlsleben. Diese verlaufen auf verschiedenen Ebenen der angeborenen Affekte und erworbenen Emotionen und beeinflussen, manchmal sehr stark, auch die inhaltliche Seite des Gesprochenen. Mit Bezug auf die emotionellen Äußerungen (in der Stimme und auch in der Sprache) kommen die individuellen Eigenschaften zur Geltung.

* Cf. the journal *SPFFBU*, quoted on p. 6.

SKOUPIL: Gibt es eine Möglichkeit, auf der Grundlage dieser Hypothese eine Typologie der Kinder (im psychologischen Sinne) aufzubauen?

ŠTEMBERGOVÁ: Ich arbeite in der literar-dramatischen Erziehung der Kinder. Ich habe bemerkt, daß die Gefühle und Emotionen, die sich in gewöhnlicher Kindersprache ganz genau in Modulationen spiegeln, plötzlich aufhören, wenn das Kind gezwungen wird, recht gut grammatisch und orthoepisch zu sprechen. Das Kind fühlt sich gar so viel gebunden, daß es sich war richtig, aber flach und neutral ausdrückt. Wenn ich ihm erlaube, sich ein bischen gesprächlich auszudrücken, kommen die Modulationen wieder her und die Sprache belebt wieder.

DÉVELOPPEMENT DES CONSTRUCTIONS DE COMPLÉMENT D'OBJET ET DE LIEU DANS LE LANGAGE DES ENFANTS SERBOCROATES ET HONGROIS

P. VLAHOVIĆ — M. MIKEŠ — L. DEZSŐ

(Novi Sad-Budapest)

Ce travail n'est qu'une partie de recherches sur le syntagme verbal dans le langage enfantin à l'âge précoce, de 1;2 ou 1;3 à 1;7 ou 1;8. Notre but est de déterminer la fréquence des constructions de complément d'objet et de celles de complément de lieu par rapport à la fréquence de tous les autres énoncés enregistrés dans la dite période, ainsi que la fréquence des constructions de complément d'objet par rapport à la fréquence des constructions de complément de lieu. Cette analyse doit contribuer aux recherches sur la genèse de la composante de complément d'objet et de celle de complément de lieu dans la compétence linguistique de l'enfant.

Ces constructions à l'âge examiné nous offrent un riche matériel pour l'observation de l'ordre de mots. Pourtant, cet âge ne nous donne pas assez de matériel pour l'observation de la déclinaison et de l'emploi de la préposition ou de la postposition. Ce qu'on pouvait en tirer est déjà analysé dans notre travail sur le développement des catégories grammaticales dans le langage enfantin.*

Pour l'analyse des constructions de complément d'objet et de lieu nous nous sommes servis du matériel linguistique des quatre sujets examinés dans notre ouvrage "Sentence Programming Span" (D_1, né en 1958, D_2, né en 1961, D_3, né en 1962 et D_4, né en 1965). Ces enfants vivent à Novi Sad (Yougoslavie). D_1 et

* M. Mikeš - P. Vlahović, *Razvoj gramatičkih kategorija u dečjem govoru*: *Prilozi proučavanju jezika* II (Novi Sad, 1966).

D_2 sont enfants bilingues: hongrois et serbocroate; D_3 et D_4 sont enfant monolingues: serbocroate.)

L'analyse quantitative comprend une période d'observation intensive du développement des constructions de complément d'objet et de lieu dans le langage de D_1 — enfant bilingue — et dans celui de D_4, enfant monolingue. Cette période concerne l'étape évolutive à partir de l'apparition des premiers énoncés à deux membres jusqu'à l'étape prétransformationnelle de la phrase composée. Comme nous l'avons déjà dit dans notre communication "Sentence Programming Span" (cf. p.166), cette période embrasse 3 mois chez D_1 (de 1;2 à 1;4) et 6 mois chez D_4 (de 1;3 à 1;8). L'évolution des constructions de complément d'objet et de lieu est démontrée en 6 sous-étapes de demi-mois chez D_1 et d'un mois chez D_4.

Les données obtenues montιent que les constructions de complément d'objet et de lieu, c'est-à-dire, les énoncés dont l'interprétation exige l'introduction de la composante de complément d'objet et de lieu représentent une grande partie d'énoncés enregistrés à l'âge examiné. Voir O/L (1).

Dans le langage de D_4 à l'âge de 1;3 ces structures font la moitié des énoncés notés; à l'âge de 1;4 leur pourcentage augmente pour atteindre son point culminant à l'âge de 1;5 ($\frac{3}{4}$ de tous les énoncés). A l'âge de 1;6 le nombre de ces constructions diminue (47%) et à l'âge de 1;7 et 1;8 augmente une seconde fois, dépassant la moitié de tous les énoncés.

Dans le langage de D_1 les tendances quantitatives sont semblables à celles du langage de D_4. Il y a quand même quelques différences: au début les constructions de complément d'objet et de lieu font 62% de tous les énoncés notés, alors qu'aux étapes suivantes leur pourcentage baisse jusqu'à 35% pour atteindre à la fin de la période 51% de tous les énoncés.

L'analyse comparative des constructions de complément d'objet et de celles de lieu dans le langage de D_4 a montré que les constructions de complément d'objet sont plus nombreuses au cours de toute la période examinée: à l'âge de 1;3 le rapport est 36%: 14%; à l'âge de 1;4 la différence diminue et fait 28%: 24%, alors qu'à l'âge de 1;5 les constructions de complément d'objet sont très fréquentes et les constructions de complément de lieu très rares (67%: 8%); à l'âge de 1;6 le nombre des premières diminue (34%) et le nombre des secondes augmente (13%); à l'âge de 1;7 et 1;8 le pourcentage des constructions de complément d'objet monte jusqu'à 40% et celui des constructions de complément de lieu reste presque au même niveau.

La même analyse du matériel de D_1 montre au total que les constructions de complément d'objet sont plus nombreuses que les constructions de complément de lieu, comme chez D_4. Pourtant les différences quantitatives sont moins remarquables.

Étant donné qu'au début de l'acquisition des constructions l'enfant n'exprime qu'un constituent du syntagme verbal, et seulement plus tard les deux, nous distinguons dans l'analyse trois types de constructions:

(1) le verbe exprimé; le complément sous-entendu par la situation;

(2) le complément exprimé; le verbe sous-entendu par la situation;

(3) le verbe et le complément exprimés.

Les constructions de complément d'objet, dont nous avons déjà constaté le grand nombre dans le répertoire des énoncés enfantins, sont très importantes dans la vie de l'enfant qui s'en sert pour exprimer ses besoins vitaux en communiquant avec son milieu — en montrant ou en demandant la nourriture, les jouets et d'autres objets, soit qu'il désigne l'action, l'objet ou l'action et l'objet.

O1 est le type de construction par laquelle l'enfant n'exprime que le verbe; il désigne l'action mais le cadre de l'énoncé dans une situation définie exige absolument l'introduction de la composante d'objet dans l'interpretation de la phrase:

(1) *bu-bu (ljubi)* 1;3
D_4 embrasse un jouet.

O2 représente la construction par laquelle l'enfant n'exprime que l'objet; il désigne l'objet qu'il veut que quelqu'un lui donne ou qu'il veut prendre lui-même:

(2) *bibe (cipele)* 1;3
D_4 donne ses souliers à sa mère pour qu'elle les lui chausse.

O3 est la construction par laquelle l'enfant exprime le verbe et l'objet:

(3) *ode-ode # daj-daj (vode daj)* 1;4
D_4 demande de l'eau.

Les constructions de complément de lieu représentent les énoncés contenant la composante de lieu, par ex.: l'enfant fait savoir à son milieu où se trouve lui-même, quelqu'un d'autre ou quelque chose.

L1 est la construction par laquelle l'enfant ne désigne que l'action mais le cadre de l'énoncé dans une situation définie exige absolument l'introduction de la composante de lieu dans l'interpretation de la phrase:

(4) *tata* 1;4
D_4 le dit en montrant la porte de la chambre de son père.

L2 est la construction par laquelle l'enfant ne désigne que l'endroit où l'action se déroule ou doit se dérouler:

(5) *koja (kola)* 1;4
D_4 le dit désirant s'asseoir dans sa voiture.

L3 est la construction par laquelle l'enfant exprime le verbe et le complément de lieu:

(6) *dojo-dojo # uti (dole ustane)* 1;6
D_4 veut descendre du divan.

Le développement des constructions O1, O2, O3 et L1, L2, L3 est suivi et analysé dans la dite période de 6 mois, ce qui nous a donné les résultats contenus dans le tableau statistique O/L (2).

Dans le langage de D_4, la construction O1 montre la tendance d'une baisse légère à partir de 1;3, où elle apparaît avec 29%, jusqu'à 1;7 et à 1;8 la tendance d'une baisse très forte (3%).

La construction O2 apparaît avec le même pourcentage que O1; à l'âge de 1;5 elle atteint son point culminant (55%), après quoi elle baisse, juste à la période où le pourcentage de la construction O3 augmente.

Tandis que les trois types de constructions de complément d'objet apparaissent à tous les âges, on ne trouve que deux types de constructions de complément de lieu au même âge. C'est seulement à 1;7 et 1;8 qu'on a enregistré les trois types.

Il faut souligner que les constructions O1 et L1 disparaissent presque complètement à la fin de la période de 6 mois. Les constructions les plus complexes O3 et L3 sont les plus fréquentes à l'âge de 1;8.

Dans le langage de D_1 la construction O1 atteint son point culminant à l'âge de 1;3 et disparaît presque complètement à la fin de 1;4.

O2 apparaît avec un pourcentage élevé, puis baisse, mais au bout de la période garde toujours 22%.

O3 est la plus forte des constructions précédentes. Elle commence par 19% à l'âge de 1;2, après quoi elle augmente et atteint à la fin de la période 31%.

Des constructions de compléments de lieu, la moins fréquente et la moins stable, c'est la construction L1.

Quant aux constructions L2 et L3 la première est plus fréquente jusqu'à l'âge de 1;3 alors que l'autre accroît son pourcentage à partir de cet âge. A 1;4 O3 et L3 sont les plus fréquentes.

Vu que l'enfant déjà à l'âge précoce non seulement imite le langage de son milieu mais aussi crée tout seul des constructions, celles-là diffèrent des constructions du langage des adultes. L'analyse de la période examinée nous a rendu intéressant le phénomène de la permutation. Dans l'ouvrage qui traite les questions théoriques et méthodologiques des recherches syntaxiques du langage enfantin (*Teorijsko-metodološka pitanja izučavanja sintakse dečjeg govora*; *Prilozi proučavanju jezika* [Novi Sad, 1970]), L. Deže et P. Vlahović ont discuté sur la place du complément d'objet et du verbe. Pourtant l'analyse de la période indiquée nous a permis de faire quelques observations nouvelles.

(Voir Données statistiques sur la permutation O3 D_3 et D_4)

Dans des constructions de complément d'objet jusqu'à 1,6 chez D_4 l'ordre de mots est OV et ce n'est qu'à partir de cet âge qu'apparaît l'ordre de mots VO. Pendant que les constructions avec OV disparaissent peu à peu jusqu'à la fin de 1;8, le nombre de constructions avec VO accroît rapidement. L'ordre de mots OV à tous les âges apparaît exclusivement dans les constructions à deux mots avec l'intonation discontinue de la phrase (phase 2 de SPS):

(7) *jatuk # mete-mete*
D_4 veut que sa mère mette le coussin dans sa voiture (1;5).
(*jatuk jastuk* sc 'coussin', *mete* sc 'elle met').

Au contraire, on peut trouver l'ordre de mots VO dans des constructions à deux mots avec l'intonation discontinue ainsi que dans des constructions à deux mots avec l'intonation continue de la phrase (phase 3 de SPS). Ces dernières sont dominantes à l'âge 1;7:

(8) *kine daje*
D_4 veut enlever ses sandales (1;6).

(*kine skine* sc 'elle enlève', *daje sandale* sc 'sandales')

Le langage de D_3, pourtant, reflète l'imitation du langage des adultes, c'est-à-dire l'ordre de mots typique en serbocroate VO domine presqu'absolument dans des constructions de complément d'objet direct. (Voir Données statistiques sur la permutation O3 D_1 et D_2)

Les tendances du cours évolutif de la permutation chez les enfants bilingues D_1 et D_2 diffère de celles chez les enfants monolingues, dont nous venons de parler. Il faut mentionner qu'en hongrois il y a, à part l'ordre de mots VO, un autre ordre de mots également typiques, OV. Ce phénomène a de l'importance dans le développement des constructions de complément d'objet chez nos enfants bilingues:

(9) *tüsü # add ide*
D_2 demande le peigne (1;7).
tüsü fésü h 'peigne', *add ide* h 'donne'

(10) *add ide cipelu*
D_2 demande son soulier (1;7).
cipela sc 'soulier'

Les deux types d'ordre de mots montrent des tendences à peu près parallèles, mais OV est d'une fréquence presque toujours inférieure à celle de VO. Ce qui est peut être le plus significatif, c'est que dans le développement chez D_1, vers la fin de la période examinée, la courbe VO montre une petite baisse, alors que la courbe OV accroît rapidement.

L'analyse plus détaillée de la fin de la période observée (sous-période VI) chez nos enfants bilingues nous permet de constater le suivant:

Les constructions de complément d'objet serbocroates, qui n'apparaissent dans le répertoire de D_1 qu'assez tard (sous-période V), contiennent l'ordre de mots VO. Les nouvelles combinaisons hongroises et serbocroates à la fin de la période examinée (sous-période VI) sont générées avec l'ordre de mots OV. Dans la même période D_4, enfant monolingue, abandonne l'ordre de mots OV, en le substituant par VO. Donc, le type d'ordre de mots OV nous semble plus important pour la génération des constructions de complément d'objet chez tous les enfants. Une plus grande fréquence de l'emploi de l'ordre de mots VO chez nos enfants monolingues ainsi que la tendance croissante de l'emploi de OV chez nos enfants bilingues, nous montre, quand même, que l'influence du langage du milieu linguistique joue un rôle important déjà à l'âge précoce de l'enfant.

ANNEXE

Liste des exemples

(1) *bu-bu*
D_4 embrasse un jouet (1;3)
bu-bu (ljubi) sc 'embrasse'

(2) *bibe*
D_4 donne ses souliers à sa mère pour qu'elle les lui mette (1;3)
bibe (cipele) sc 'souliers'

(3) *ode-ode # daj-daj*
D_4 demande de l'eau (1;4)
ode-ode # daj-daj (vode daj) sc 'donne-moi de l'eau'

(4) *tata*
D_4 le dit montrant la porte de la chambre de son père (1;4)
tata sc 'papa'

(5) *koja*
D_4 le dit désirant s'asseoir dans sa voiture (1;4)
koja (kola) sc 'la voiture d'enfant'

(6) *dojo-dojo # uti*
D_4 veut descendre du divan (1;6)
dojo-dojo # uti (dole) sc 'en bas', *(ustane)* sc 'descend'

(7) *jatuk # mete-mete*
D_4 veut que sa mère mette le coussin dans la voiture (1;5)
jatuk (jastuk) sc 'coussin', *mete* sc 'elle met'

(8) *kine daje*
D_4 veut enlever ses sandales (1;6)
kine (skine) sc 'elle enlève', *daje (sandale)* sc 'sandales'

(9) *tüsü # add ide*
D_2 demande le peigne (1;7). *tüsü (fésü)* h 'peigne', *add ide* h 'donne'

(10) *add ide cipelu*
D_2 demande son soulier (1;7) *cipela* sc 'soulier'

Données statistiques O/L (1) D_4 et D_1

D_4					D_1				
Âge	O	L	X	Total	Âge	O	L	X	Total
1;3	5	2	7	14	1;2 (I)	9	7	10	26
	36%	14%	50%			35%	27%	38%	
1;4	6	5	10	21	1;2 (II)	11	6	32	49
	28%	24%	48%			23%	12%	65%	
1;5	16	2	6	24	1;3 (I)	5	9	24	38
	67%	8%	25%			13%	24%	63%	

D4				
Age	O	L	X	Total
1;6	16	6	25	47
	34%	13%	53%	
1;7	64	18	56	138
	46%	13%	41%	
1;8	47	18	48	113
	42%	15%	43%	

D1				
Age	O	L	X	Total
1;3 (II)	37	23	84	144
	26%	16%	58%	
1;4 (I)	42	38	139	219
	19%	17%	64%	
1;4 (II)	52	44	92	188
	28%	23%	49%	

Données statistiques O/L D4

Âge	O 1	O 2	O 3	L 1	L 2	L 3	Total
1;3	2	2	1		1	1	7
	29%	29%	14%		14%	14%	
1;4	3	2	1	2	3		11
	27%	18%	10%	18%	27%		
1;5	4	10	2	1	1		18
	22%	55%	11%	6%	6%		
1;6	4	7	5		3	3	22
	18%	32%	22%		14%	14%	
1;7	14	23	27	4	5	9	82
	17%	28%	33%	5%	6%	11%	
1;8	2	14	30	1	4	14	65
	3%	22%	46%	2%	6%	23%	

Données statistiques O/L (2) D1

Âge	O 1	O 2	O 3	L 1	L 2	L 3	Total
1;2 (I)	1	5	3	1	6		16
	6%	31%	19%	6%	38%		
1;2 (II)	1	1	9		4	2	17
	6%	6%	52%		24%	12%	
1;3 (I)	3		2	2	5	2	14
	22%		14%	14%	36%	14%	
1;3 (II)	5	5	27	1	7	15	60
	8%	8%	45%	2%	12%	25%	
1;4 (I)	9	10	23		5	33	80
	11%	13%	29%		6%	41%	
1;4 (II)	1	21	30		15	29	96
	1%	22%	31%		16%	30%	

Données statistiques sur la permutation O 3

Sous-périodes	D_1		D_2		D_3		D_4	
	VO	OV	VO	OV	VO	OV	VO	OV
I	2	1	2	2	0	0	0	1
II	8	1	4	5	0	0	0	1
III	1	1	9	1	0	0	0	2
IV	22	5	6	0	0	0	3	3
V	21	2	7	0	20	1	23	3
VI	19	11	41	13	74	1	29	1

DISORDERS OF SPEECH IN CHILD APHASIA

MARIA ZARĘBINA*

(Kraków)

The paper deals with disorders in the phonetic and phonological systems of three children afflicted with child aphasia (congenital aphasia). The children (two boys of 8 and 11, and a girl of 9) could not speak in their early infancy and have now been learning to speak in a special school.

The eleven year-old boy, T.Z., has been observed for one month. He can only repeat certain syllables when requested; sometimes he can answer a question with one word, but unsollicited does not speak at all. He pronounces open syllables mainly, for example *tu*, *to*, *ńe*, *to-ma (Tomasz)*, *to-me (Tomek)*, *ma*, *i̯a*, but also *pan*, *bam (bat)*, *tam*, which are closed with a sonant; on one occasion he even pronounced *tak*, a syllable closed with a plosive. He divides words longer than one syllable into syllables, for instance *te-ka (torebka)*, *i̯i-gu̯a*. The boy's phonological system is incomplete; his inventory of phonemes embraces all the oral vowels (no nasal vowels) and the following consonants: the plosives *p*, *b*, *t*, *k*, *g*, the spirants *s*, *z*, *š*, *ś*, *ź*, *x*, the affricates *c*, *č*, the sonants *m*, *n*, *ń*, *l*, as well as *j* /i̯/ and *ł* /u̯/. The system, then, lacks *d*, *ž*, *ʒ*, *ǯ*, and *r*. Repeating syllables after me the boy substitutes *r* with *l*. Also the substitution of voiceless phonemes for voiced ones can be observed: the boy repeats *da* as *ta*, *baba* as *pa-pa*, *dzisiaj* as *ći-śa*, *ga* as *ka*. One also finds in his speech simplifications in the sound and syllable

* *Not present in person*

structure of words, for example *ma (mam)*, *ta (tak)*, *o-ka (okulary)*, *i̯e-i̯a (jedenaście)*, *ću-te (klucze)* (with metathesis and phoneme subsitution*)*, *u-va (ołówek)*.

Though the boy speaks so little he can understand very much; he can indicate four objects named by me all at once, he can count to 10 in writing (changing the given order of numerals), he can also write a few letters and syllables, but does not know the whole alphabet. He can communicate many things by means of gestures, for instance he has been able to let me know in this way that he was out for a walk with a group of children and heard the bugle-call from Our Lady's Church at 11 o'clock.

The girl, K.S., 9 years old, repeats words after me and answers questions, usually with one word (in most cases a noun in the nominative), which she does not divide into syllables; there are also answers of more than one word, e.g. /How long have you been here?/ *c'eć'i lok (trzeci rok)*, /What colour is this?/ *Iagby talny (Jakby czarny)*, and even sentences /What does your mother do?/ *Myi̯e-okna*; *Psyi̯ezʒa-do mné-mama*. These sentences are divided into words, though occasionally a sentence without such division may occur, e.g. *Kotek tak pije*.

The girl has mastered the phonological system, EXCEPT THE PHONEME *r*, which is CONSISTENTLY REPLACED, while the other substitutions to be mentioned exist by the side of correct realizations. The phoneme *r* is most often substituted by *l*, e.g. in *Klystyna*, *dluǵi (drugi)*, *s'ostla (siostra)*, *blata (brata)*, *tovalovy (towarowy)*, *ḱelońis'ka (kierowniczka)*, *tolba (torba)*, *kaltka (kartka)*. Exceptionally *r* may also be substituted by *ł*: *su̯ebu̯e (srebrne)*, *su̯oda (środa)*, or even by *j*: *sebi̯e (srebrne)*, *nai̯ty (narty)*.

There are SUBSTITUTIONS within the OPPOSITION VOICELESS: VOICED, with the voiceless sounds replacing the voiced ones, e.g.: *koloncy (gorący)*, *kuz'ik (guzik)*, *nofe (nowe)*, *fidok (widok)*, *f'izǫ (widzą)*, *vasa (waza)*, *pav́imy śe (bawimy się)*, *sapomńau̯am (zapomniałam)*, *su̯ote (złote)*, *slobi̯u̯am (zrobiłam)*, *tlaṕina (drabina)*, *pai̯ka (bajka)*, *selo (zero)*.

There are SUBSTITUTIONS within oppositions based on the

MANNER OF ARTICULATION. Thus, plosives replace affricates, e.g. *talne (czarne)*, *pšenta (przędza)*, *pšentony (przędzony)*, *śf'eťiuo (świeciło)*, *denny (dzienny)*, *tapa (capa)*, *dlut'ik (drucik)*, *tajńik (czajnik)*, *to (co)*, *debendeśont (dziewięćdziesiąt)*; affricates are substituted by spirants, e.g. *śtely (cztery)*, *uĺisny (uliczny)*, but on the other hand they may also replace spirants: *kledenc (kredens)*, *jazembina (jarzębina)*, *ćveže (zwierzę)* (though the last word may be an example of dissimilation).

The following THREE CONSONANT SERIES ARE MIXED: *s z c ʒ*; *š ž č ǯ*; and *ś ź ć ʒ́*. In fact, only two of them exist: the dentals and the mid-consonants; the alveolar series is missing. Only occasionally, while repeating single words after me, the girl produces e.g. *šafa*, but this does not happen when she answers questions. As the mid-consonants are often realized as dentals (with slight palatalization instead of full palatalization), the alveolar consonants are sporadically replaced by the dentals (palatalized), the three series are mixed. Thus, instead and by the side of hard dentals, e.g. *Klystyna Symcak (Krystyna Szymczak)*, *tostaua (dostała)*, we observe palatalized dentals, e.g. *vos'ku (w wojsku)*, *s'tona (strona)*, *s'ostla (siostra)*, *wys'oko (wysoko)*, *c'o (co)*; instead of alveolar consonants we get dentals (hard or palatalized), e.g. *nozycḱi (nożyczki)*, *kalendas (kalendarz)*, *saf'ilovy (szafirowy)*, *suḿi (szumi)*, *jasempin (jarzębin)*, *kluce (klucze)*, *sanuj (szanuj)*, *tons'ka (książka)*, *puasc (płaszcz)*, *ps'i-moc'ovac' (przymocować)*, *fcola (wczoraj)*; instead of mid-consonants we get dentals, e.g. *seʒec (siedzieć)*, *naʒevac (nadziewać)*, *se (się)*, *dvanasce (dwanaście)*, or palatalized dentals, e.g. *Maugos'a (Małgosia)*, *Danus'a (Danusia)*, *Maǯa (Madzia)*, *ṕejńc' (pięć)*, *ʒ'es'eńć (dziesięć)* (by the side of the correct forms: *guźik, ńići, śenńik)*. Assimilation is frequent within the three consonant series, e.g. *lysovac (rysować)*, *sesyt (zeszyt)*, *susyc (suszyć)*. Also dissimilation frequently occurs, e.g. *dvaǯesća (dwadzieścia)*, *ṕesćonek (pierścionek)*.

Apart from the above substitutions, there are occasional disturbances in the girl's pronunciation of particular sounds. For example, there exists a TWO-PHASE VOICED-VOICELESS OR VOICELESS-VOICED articulation, e.g. *s^{z}apau̯ (zapal)*, *zap^{b}a (zaba)*, *g^{k}ub́p'i (gubi)*, *nog^{k}a (noga)*, *b^{p}av́f'ić śe (bawić się)*, *daleg^{k}o (daleko)*. Besides, the INTERDENTAL *Θ* appears here and there, e.g. *Θalne (czarne)*, *Θo (co)*, *Θemu (czemu)*, *plaΘa (praca)*. This sound is one way of expressing the affricates; their dental or alveolar quality is of secondary importance, since the alveolar consonants can be replaced by the corresponding dentals. Another way of articulating the AFFRICATES is their articulation as PLOSIVE + SPIRANT, e.g. *kluts (klucz)*, *tsap (cap)*, *tśixo (cicho)*. Conversely, the sequence plosive + spirant may give an affricate, e.g. *ćoska (książka)*, *cesu̯o (krzeslo)*, (k + *ś* ⩾ *t* + *ś*, there is assimilation as regards the place of articulation, *t* + *ś* ⩾ *ć*; *k* + *š* ⩾ *t* + *š* ⩾ *č*; *č* is subsequently replaced by c).

In addition to those phonetic peculiarities of the girl's speech, we also observe in it SIMPLIFICATION OF CONSONANTAL GROUPS, e.g. *of'eʒ̇iny (odwiedziny)*, *zeb́eń (grzebień)*, *pou̯oga (podłoga)*, *sufada (szuflada)*, *podencny (podręczny)*, *vozʒ̇e (gwoździe)*, *la (dla)*, *ńasto (gniazdo)*, *stona (strona)*, *napastek (naparstek)*; ASSIMILATION, e.g. *u̯atu̯e (łatwe)*, *tytlyna (cytryna)*, *sapasy (zapasy)*, *telata (cerata)*, especially the above-discussed assimilations within the three consonantal series; sometimes DISSIMILATION, e.g. *su̯ot'isće* or *su̯oćiśte (złociście)*, *zńenca ([do] znięcia)*; ANTICIPATION, e.g. *cyklḱel (cyrkiel)*, *fultelko (futerko)*, *klendenc (kredens)*, *kolpelta (koperta)*; METATHESIS, e.g. *fulto (futro)*, *klenel (kelner)*, *xyftľivy (chwytliwy)*.

The 8 year-old boy, A.B., answers questions with developed clauses, e.g. *Ja się nazywa (!) A.B.*; *Była niebieska woda w stawie*; *Oglądały wodę i ryby*; *My się bawimy z dziećmi*. Complex sentences also occur, e.g. *Jest, tylko nie wiem*, as well as subordinate clauses standing alone when the main clause is contained in the question, e.g. *(Co to jest koza?) Co ma duze rogi*. The syntax is impoverished to some extent (for instance, there is no infinitive, and the oblique cases are replaced by the nominative), but I shall not enter into details here. Both this boy and the girl can read and write, but they make mistakes, the number of which increases as we pass from dictation or copying to autonomous writing.

Of the three children, this boy's pronunciation is the best, but deviations from the norm appear also in his case. The OPPOSITION VOICED VS. VOICELESS CONSONANTS IS NOT FIXED, and mutual substitutions happen within it, that is voiceless consonants occur instead of voiced ones, e.g. *nazyfam (nazywam)*, *Sakopanym (w Zakopanem)*, *sapau̯ki (zapałki)*, *pai̯ke (bajke)*, and voiced consonants occur instead of voiceless ones: *Vranek (Franek)*, *flagońik (flakonik)*, *suv́it (sufit)*, *wybrysk (wyprysk)*. The series *s z c ʒ* and *ś ź ć ʒ́* are used correctly, but the alveolar series is now and then substituted by the dental series. Thus, we have correct forms like *i̯ežyk (Jerzyk)*, *žaba (zaba)*, *čapka (czapka)*, *šafa (szafa)*, but also incorrect forms like *ctery (cztery)*, *i̯anus (Janusz)*, *kalendas (kalendarz)*, *cemu (czemu)*, *sanui̯ (szanuj)*. When the boy reads he pronounces words correctly, e.g. *šust (szust!)*, *pšygody (przygody)*. Nasal vowels before spirants and in word-final position are pronounced as the groups ou̯, eu̯, or o^{m}, e.g. *Bou̯xur (Bąchór)* [a name], *vou̯ska (wiązka)*, *negeu̯ (nogę)*, *toneu̯ (tonę)*, *za žeckom (za rzeczką)*, *viʒom (widzą)*. In other positions we sometimes observe lack of nasalization, e.g. *rosnoć (rosnąć)*, *podrecny (podręczny)*, though this can also be interpreted as simplification of the consonantal group created by the splitting of the nasal *(podrenčny = podrečny)*. SIMPLIFICATION OF CONSONANTAL GROUPS occurs from time to time, e.g. *ćina (ścina)*, *sevetka (serwetka)*, *naraṕany (nakrapiany)*, *fytlivy (chwytliwy)*. *Zonek (dzwonek)* is an interesting example: in this case the simplification of the group must have been preceded by an articulation of the affricate as plosive + spirant *(dzvonek* for *ʒvonek)*. ASSIMILATION occurs frequently, e.g. *sostra (siostra)*, *ležeš (lezeć)*, *kororovy (kolorowy)*, *okurare (okulary)*, *nozyski (nozyczki)*, *šešč (sześć)*, *pisas (pisać)*, *pšenža (przędza)*. DISSIMILATION is more rare: *ʒ́ev́enc (dziewięć)*, *ṕenc (pięć)*, *ḿeʒany (miedziany)*, *seʒ́eć (siedzieć)*, *nozyčḱi (nożyczki)*. METATHESIS is also infrequent: *liloneum (linoleum)*. Here a peculiar example is again furnished by the form *zdvonek*: metathesis is preceded by an articulation of the affricate as plosive + spirant *(ʒvonek ⇒ dzvonek ⇒ zdvonek)*.

All the three cases of child aphasia described here represent the same kind of aphasia at different stages. Disorders appear mainly in the phonetic and phonological systems. Many of the disorders are analogous to those occurring during the normal child's speech acquisition. Examples of substitutions within oppositions based on the manner of articulation, mixing the three consonant series, replacement of *r* by *l*, *ł*, or *j*, lack of nasal vowels, simplification of consonantal groups, special types of assimilation, dissimilation, metathesis, and anticipation have all been attested in studies of speech acquisition in Polish and foreign children.[1] The same phenomena occur in motor speech aphasia, which in linguistic terms is described as aphasia of contiguity relations in speech production.[2] Child aphasia is distinguished from common aphasia by the fact that what happens in the latter is the loss of speech previously acquired and mastered, while in the former the child has never spoken before.[3] Apart from phonetic and phonemic phenomena common to child aphasia, to the speech of the child who is just learning to speak, and to motor speech aphasia, the cases described above present phenomena which do not occur in the normal child's speech acquisition process. In all the three cases we observe disorders in the consonantal opposition voiceless-voiced; disorders of this kind appear in the speech of almost all motor aphasics, but Polish children master this opposition very early.[4] Some other articulations faulty from the point of view of the Polish phonetic system are still to be mentioned. These are: the two-phase (voiceless-voiced and voiced-voiceless) articulations, Θ, and plosive + spirant instead of an affricate, and vice versa. These and similar articulations are familiar to me from the observation of people

[1] L. Kaczmarek, *Rozwój mowy dziecka* (Poznań, 1953); K. Ohnesorg, *Fonetická studie o dětské řeči* (Praha, 1949); A. Grégoire, *L'apprentissage du langage* (Liège-Paris, 1937).

[2] R. Jakobson and M. Halle, *Fundamentals of Language* (The Hague, 1956); I. Cyran, "Afazja w zakresie systemu fonologicznego", *Lódzkie Towarzystwo Naukowe*: *Rozprawy Jezykowe* XV (1969), 177-87; M. Zarębina, "Afazja jako zagadnienie jezykoznawcze", *Biuletyn Fonograficzny* X (1969), 129-39.

[3] M. Maruszewski, *Afazja*: *Zagadnienia teorii i terapii* (Warszawa, 1966).

[4] J. T. Kania "Narastanie systemu fonologicznego w mowie dziecka i jego rozpad w afazji", *Folia Societatis Scientiarum Lublinensis*, sectio A-D 9/10, suppl. (1969-70), 55-60.

suffering from motor aphasia with light apraxia. It is, then, possible to demonstrate the existence of phonetic and phonological analogies between child aphasia, the speech acquisition of the normal child, motor aphasia and apraxia.

CLOSING SESSION

M. KLOSTER-JENSEN

Mr. President, Dear Colleagues,

There is no specialization without complication, no complication without the need for new approaches and departures. This fact has been strongly recalled to our minds during the symposium which we are about to close. The study of child language rests on a range of sciences. It would be an impossible task to make out, at the present stage of research, anything in the direction of a general programme of investigation.

We have been happy to learn that the papers presented are to be published. Yet we are in a position to add another important feature to this symposium. I am referring to the proposal made by Professor Ohnesorg to establish university chairs in Pedolinguistics. As you have unanimously approved of this plan here and now, I shall be glad to report to Unesco within three days about this motion on our part. It is vigorously supported by your intention to create an International Society for the Study of Child Language, whose birth we shall soon witness.

EPILOGUS

KAREL OHNESORG

Sodales carissimi!

Iam opus exegimus et nunc ratio colloquii nostri reddenda est. In congressu nostro, sui generis primo, luce clarius apparuit, quam multae multiplicesque essent quaestiones solvendae, cum studio linguae infantium operam dare vellemus. Non mirum igitur, quod investigatores inveniuntur, qui in operibus suis praecipue has quaestiones solvere conati, originem disciplinae novae, paedolinguisticae, immo vero autem paedophoneticae quoque, vaticinati sint. Quae spes eos non fefellit. Etenim colloquium nostrum permulta attulit argumenta, quae hanc opinionem confirmarent, et simul fundamenta harum disciplinarum novarum adeo corroboravit, ut nemo postea de earum loco in numero scientiarum dubitare possit. Ergo crescant, augeant, sint semper in flore!

Necesse autem est, ut postero tempore plures auctores, diversas scientiae disciplinas tractantes, ex composito communi munere fungantur ac hoc modo totis quaestionibus ad nostram materiam pertinentibus satisfaciant. Q.F.F.F.Q.S.

Gratias ago vobis omnibus, qui lectiones habuistis, vel munere praesidis suscepto conventibus nostris praefuistis, disputationibus interfuistis, colloquiis adfuistis. Exopto vobis in novis investigationibus ac in operibus futuris successus plenissimos, salvere vos iubeo.

Quoniam autem omnes quaestiones in nostro colloquio propositae iuste recteque pertractatae sint, declaro symposium nostrum perfectum, perpetratum, clausum. VALETE!

COLLOQUII PARTICIPES

ad Kal. Octobres inscripti

Bartoš, *Lubomír*, Doc. Dr., Šumavská 30, Brno ČSSR
Becker, *Ruth*, Prof. Dr. habil., Schumannstrasse 20/21, Berlin DDR
Bederová, *Drahomíra*, prom.ped., Legionárska 10, Bratislava ČSSR
Borel-Maisonny, *Suzanne*, Présidente de C.F.A.P., 10, Rue de l'Arrivée, Paris 15e France
Buchtelová, *Růžena*, Dr., Senegalská 635, Praha 6 ČSSR
Coutinho, *Dick*, rédacteur, 5 Herderstraat, P.O. Box 1132, The Hague Nederland
Dezsö, *Lázsló*, Dr., Szalay utca 10-14, Budapest V Hungaria
Dvončová, *Jana*, Dr., Rauchova 24, Bratislava ČSSR
Hartmann, *Hans*, Prof. Dr., Von-Mele-Park 6, Hamburg 13 BRD
Chmura-Klekotowa, *Maria*, Dr., ul.Nowolipie 12., m.ll, Warszawa Polska
Janota, *Přemysl*, Doc. Dr., Hrusická 4, Praha 4 ČSSR
Janotová, *Naděžda*, prom.ped., Hrusická 4, Praha 4 ČSSR
Kaczmarek, *Leon*, Prof. Dr. (*cum uxore Zofia*), ul. Nowotki 10, Lublin Polska
Kania, *Józef*, Dr., ul. Nowotki 10, Lublin Polska
Kaper, *Willem*, Wouwermanstr. 39hs, Amsterdam - OZ Nederland
Kloster-Jensen, *Martin*, Prof. Dr., Phonetics Institut, Bergen Norge
Kurcz-Piesowicz, *Ida*, Dr., Instytut psychologii, Warszawa 64 Polska
Kyselá, *Jaroslava*, Dr., Radlas 16, Brno ČSSR
Lejska, *Vladislav*, Dr., foniatrické odd. nemocnice, Kyjov ČSSR

Lindner, *Gerhart*, Prof. Dr., Bleckmannweg 1, Berlin-Lichtenberg DDR
Liška, *Jozef*, Dr., Narcisová 2, Košice ČSSR
Meggyes, *Klára*, *S.*, Prof., Szalay utca 10, Budapest V Hungaria
Mikeš, *Melania*, Dr., Fruškogorska A-6, u.1, st.4, Novi Sad Jugoslavija
Milan, *Mikuláš*, Dr., Štúrova 5, Bratislava ČSSR
Novák, *Zdeněk*, Dr., Mikulandská 5, Praha 1 ČSSR
Ohnesorg, *Karel*, Prof. Dr., Šumavská 36, Brno ČSSR
Oksaar, *Els*, Prof. Dr., Parkberg 20, Hamburg 66 BRD
Ondráčková, *Jana*, Dr., Nad Primaskou 15, Praha 10 ČSSR
Pačesová, *Jaroslava*, Doc. Dr., Fandrlíkova 72, Brno -Žabovřesky ČSSR
Pavlović, *Milivoj*, Prof. Dr., (*cum filia sua*) Nikola Stefanoviće 25, Beograd Jugoslavija
Pešinová, *Hana*, Dr., Mikulandská 5, Praha 1 ČSSR
Popel, *Miloš*, Ab Societatum doctarum epistulis, Engelsovo nábř. 6, Praha 2 ČSSR
Průcha, *Jan*, Dr., Mikulandská 5, Praha 1 ČSSR
Rajnov, *Vasil*, Dr., Car Šišman 41, Sofia Bulgaria
de Ridder, *Peter*, directeur, 5 Herderstraat, P.O. Box 1132, The Hague Nederland
Rigault, *André*, Prof., Mc Gill University, Montréal Canada
Rūķe-Draviņa, *Velta*, Prof. Dr., Docentbacken 3, Stockholm Sverige
Sedláčková, *Eva*, Doc. Dr., Žitná 24, Praha 2 ČSSR
Shugar, *Grace Wales*, Dr., Marszalkowska 45/49, m.24, Warszawa Polska
Schuhmacher, *W.W.*, Amanuensis, Klostergaardsvej 18, Vaerlöse Danemark
Siertsema, *Berthe*, Prof. Dr., Meander 767, Amstelveen Nederland
Skoupil, *Zdeněk*, Dr., Karmelitská 7, Praha 1 ČSSR
Sladíková, *Bianca*, prom.ped., Výchov. klinika, Opatovská cesta, Košice ČSSR
Slama-Cazacu, *Tatiana*, Prof. Doc. Dr., Rahovei 108, Bucuresti Romania

Söderbergh, Ragnhild, Doc., Borevägen 19, Sollentuna Sverige
Sovák, Miloš, Prof. Dr., Kaprova 14, Praha 1 ČSSR
Stelzig, Helmut, Doc. Dr. habil., Vulkanstr. 25, Greifswald DDR
Štembergová, Šárka, prof., U mrázovky 16, Praha 5 ČSSR
Šupáček, Ivan, Dr., Kaprova 14, Praha 1 ČSSR
Tonkova-Jampol'skaja, R.V., Prof. Dr., Frunzenskaja d.10, kv.185, Moskva G-145 SSSR
Učeň, Ivan, Dr., Legionárska 10, Bratislava ČSSR
Vajshajtlová, Marie, Dr., Mrštíkova 6, Teplice v Čechách ČSSR
Vlahović, Plemenka, Dr., Radnicki universitet, Novi Sad Jugoslavija
Vlasova, N.A., Prof. Dr., Lesnaja 63/43-131, Moskva L-55 SSSR
Vnuková, Mária, prom. ped., Legionárska 10, Bratislava ČSSR
Zarębina, Maria, Dr., Armii Ludowej 6, m. 26, Kraków Polska
Zichová, Marie, Dr., Šrobárova 36, Praha 10 ČSSR
Zlámal, Jiří, Ing., Cihlářská 17/c, Brno ČSSR

FSC
www.fsc.org
MIX
Papier | Fördert
gute Waldnutzung
FSC® C083411

Kolibri 360 GmbH
Ferdinand-Jühlke-Straße 7
99095 Erfurt, Deutschland
produktsicherheit@kolibri360.de